PRINCIPLES OF ANATOMY AND PHYSIOLOGY

Tenth Edition

Gerard J. Tortora

Bergen Community College

Sandra Reynolds Grabowski

Purdue University

John Wiley & Sons, Inc.

Cover photos: Photo of woman: ©PhotoDisc, Inc.
Photo of man: ©Ray Massey/Stone

To order books or for customer service call 1-800-CALL-WILEY (225-5945).

CONTENTS

CHAPTER 1 An Introduction to the Human Body

MULTIPLE CHOICE. Choose the one alternative that best completes the statement or answers the question.

1) Which of the following correctly lists the levels of organization from least complex to most complex?
   A) cellular, tissue, chemical, system, organ, organism
   B) chemical, cellular, tissue, organ, system, organism
   C) tissue, cellular, chemical, organ, system, organism
   D) chemical, tissue, cellular, system, organ, organism
   E) organism, system, organ, tissue, cellular, chemical

   Answer: B

2) An organ is defined as a structure that has a specific structure and is composed of two or more different types of
   A) molecules.
   B) cells.
   C) systems.
   D) tissues.
   E) membranes.

   Answer: D

3) The sum of all chemical reactions that occur in the body is known as
   A) growth.
   B) reproduction.
   C) metabolism.
   D) differentiation.
   E) responsiveness.

   Answer: C

4) Tom has been lifting weights. As a result of the physical work, his muscle cells have added proteins and become larger. Which of the following terms best describes this increase in size?
   A) metabolism
   B) growth
   C) responsiveness
   D) differentiation
   E) reproduction

   Answer: B

5) The cranial cavity is
   A) where the brain is located.
   B) lined by the meninges.
   C) part of the ventral body cavity.
   D) A and B only are correct.
   E) A, B, and C are all correct.

   Answer: D

6) A plane or section that divides an organ such that you could view an inferior surface of the section of that organ would be a(n)
   A) coronal section.
   B) medial section.
   C) sagittal section.
   D) transverse section.
   E) oblique section.

   Answer: C

7) Interstitial fluid is the fluid
   A) inside blood vessels.
   B) inside cells.
   C) between the cells in a tissue.
   D) inside lymph vessels.
   E) that is consumed as part of the diet.

   Answer: C

8) A transverse plane divides the body into
   A) superior and inferior portions.
   B) right and left halves.
   C) anterior and posterior portions.
   D) ventral and dorsal body cavities.
   E) quadrants.

   Answer: A

9) The two body systems that regulate homeostasis are the
   A) cardiovascular and respiratory systems.
   B) cardiovascular and urinary systems.
   C) cardiovascular and endocrine systems.
   D) nervous and cardiovascular systems.
   E) nervous and endocrine systems.

   Answer: E

10) The word axillary refers to the
   A) groin.
   B) armpit.
   C) upper arm.
   D) neck.
   E) back of the knee.

   Answer: B

11) The word popliteal refers to the
   A) groin.
   B) armpit.
   C) upper arm.
   D) neck.
   E) back of the knee.

   Answer: E

12) Which of the following best describes the relationship between the urinary bladder and the stomach?
   A) The urinary bladder is distal to the stomach.
   B) The urinary bladder is proximal to the stomach.
   C) The urinary bladder is inferior to the stomach.
   D) The urinary bladder is superior to the stomach.
   E) The urinary bladder is anterior to the stomach.

   Answer: C

13) Which of the following best describes the relationship between the right plantar region and the right femoral region?
   A) The right plantar region is distal to the right femoral region.
   B) The right plantar region is proximal to the right femoral region.
   C) The right plantar region is inferior to the right femoral region.
   D) The right plantar region is superior to the right femoral region.
   E) The right plantar region is anterior to the right femoral region.

   Answer: A

14) The region of the abdominopelvic cavity that is inferior and medial to the left lumbar region is the
   A) left hypochondriac region.
   B) left inguinal region.
   C) umbilical region.
   D) hypogastric region.
   E) epigastric region.

   Answer: D

15) Which of the following most correctly describes the relationship between the visceral pleura and the parietal pleura?
   A) The visceral pleura is anterior to the parietal pleura.
   B) The visceral pleura is posterior to the parietal pleura.
   C) The visceral pleura is superficial to the parietal pleura.
   D) The visceral pleura is deep to the parietal pleura.
   E) The visceral pleura is medial to the parietal pleura.

   Answer: D

16) Which of the following most correctly describes the relationship between the spine and the lungs?
   A) The spine is lateral and posterior to the lungs.
   B) The spine is medial and posterior to the lungs.
   C) The spine is lateral and anterior to the lungs.
   D) The spine is medial and posterior to the lungs.
   E) The spine is medial and deep to the lungs.

   Answer: B

17) A plane or section that divides an organ such that you would be looking at a medial surface of the section would be a(n)
   A) coronal section.
   B) horizontal section.
   C) sagittal section.
   D) transverse section.
   E) oblique section.

   Answer: C

18) Which of the following best describes the endocrine system?
   A) It regulates homeostasis by means of nerve impulses.
   B) It absorbs nutrients.
   C) It contains hair, skin, and nails.
   D) It produces blood cells that transport oxygen.
   E) It is made up of glands that secrete hormones.

   Answer: E

19) The term cephalic refers to the
   A) head.
   B) neck.
   C) back of the lower leg.
   D) chest.
   E) spinal column.

   Answer: A

20) Which of the following best describes the relationship between the ears and the tip of the nose?
   A) The ears are medial and posterior to the tip of the nose.
   B) The ears are lateral and posterior to the tip of the nose.
   C) The ears are medial and anterior to the tip of the nose.
   D) The ears are lateral and posterior to the tip of the nose.
   E) The ears are superior and medial to the tip of the nose.

   Answer: D

21) Which of the following best describes the relationship between the right shoulder and the navel?
   A) The right shoulder is inferior and medial to the navel.
   B) The right shoulder is inferior and lateral to the navel.
   C) The right shoulder is superior and medial to the navel.
   D) The right shoulder is superior and lateral to the navel.
   E) The right shoulder is superior and proximal to the navel.

   Answer: D

22) During the process of development of the skeletal system, embryonic cells, known as mesenchyme cells, may develop into either osteoblasts or chondroblasts, which, in turn, may develop into osteocytes and chondrocytes (respectively). This process is an example of
   A) growth.
   B) metabolism.
   C) differentiation.
   D) responsiveness.
   E) movement.

   Answer: C

23) The body system that distributes oxygen and nutrients to cells and carries carbon dioxide and wastes away from cells is the
   A) respiratory system.
   B) cardiovascular system
   C) endocrine system.
   D) urinary system.
   E) integumentary system.

   Answer: B

24) The system that plays the major role in regulating the volume and chemical composition of blood, eliminating wastes, and regulating fluid and electrolyte balance is the

A) respiratory system.

B) cardiovascular system.

C) endocrine system.

D) urinary system.

E) integumentary system.

Answer: D

25) Which of the following body parts would be considered ipsilateral to each other?

A) heart and diaphragm

B) right arm and right leg

C) left lung and right lung

D) collar bones and shoulder blades

E) vertebral column and digestive organs

Answer: B

26) Which of the following lines the abdominal cavity?

A) peritoenum

B) pericardium

C) pleura

D) meninges

E) mediastinum

Answer: A

27) ALL of the following are primarily studies of anatomy (as opposed to physiology) EXCEPT:

A) observing the arrangement of cells in the adrenal gland.

B) describing the process by which nerve impulses are transmitted.

C) exploring the embryonic origins of endocrine cells.

D) finding the location of the biceps femoris muscle.

E) identifying types of tissues present in the walls of the intestinal tract.

Answer: B

28) ALL of the following are primarily studies of physiology (as opposed to anatomy) EXCEPT:

A) describing the process by which glucose molecules are broken down.

B) explaining how substances are secreted from cells.

C) describing the process by which nerve impulses are transmitted.

D) identifying the types of tissues present in the walls of the intestinal tract.

E) identifying the factors that affect blood pressure.

Answer: D

29) Which of the following are considered part of the integumentary system?
   A) liver, stomach, and intestines
   B) brain and spinal cord
   C) hormone-secreting glands
   D) kidneys and urinary bladder
   E) hair, skin, and nails

   Answer: E

30) Generation of heat (thermogenesis) is a function of the
   A) integumentary system.
   B) muscular system.
   C) cardiovascular system.
   D) digestive system.
   E) nervous system.

   Answer: B

31) Osmometer cells sense changes in the concentration of blood plasma; therefore, they must be
   A) receptors.
   B) control centers.
   C) stimulators.
   D) part of the cardiovascular system.
   E) effectors.

   Answer: A

32) Osmometer cells in the brain sense an increase in the concentration of the blood plasma. They then notify the pituitary gland to release the hormone, ADH. This hormone causes the kidney to save water, which lowers the concentration of the plasma. ALL of the following are TRUE for this scenario EXCEPT:
   A) The kidney acts as an effector in this feedback loop.
   B) The osmometer cells act as receptors in this feedback loop.
   C) The stimulus in this feedback loop is an increase in the plasma concentration.
   D) The controlled condition regulated by this feedback loop is constant ADH secretion.
   E) This is an example of a negative feedback loop.

   Answer: D

33) Which of the following is an example of a positive feedback loop?

A) A neuron is stimulated, thus opening membrane channels to allow sodium ions to leak from the extracellular fluid to the intracellular fluid. This causes more membrane channels to open, thus allowing more sodium ions to enter the intracellular fluid.

B) Baroreceptors notify the brain that the blood pressure has increased. The brain then notifies the blood vessels to dilate, thus lowering the blood pressure.

C) Low levels of glucose in the blood cause the pancreas to release less insulin (a hormone that lowers blood glucose).

D) Elevated body temperature is sensed by cells in the brain. As a result, sweat is produced, and heat is lost as the water in the sweat evaporates.

E) An auto factory produces 1000 cars per week. The sales office could sell 1200 cars per week. Extra production personnel are added at the factory to meet the sales demand.

Answer: A

34) You are eating a hot fudge sundae. The pleasant taste information is sensed by your taste buds, which notify your brain. Your brain releases endorphins, which make you feel very good. You now associate the good feeling with hot fudge sundaes, so you eat another hot fudge sundae. Now you feel even better. Which of the following statements is TRUE regarding this scenario?

A) This is a negative feedback loop because two hot fudge sundaes will make you sick.

B) This is a positive feedback loop because the results make you feel good.

C) This is a negative feedback loop because you were doing something bad for your health in the first place, and the result makes the situation worse.

D) This is a positive feedback loop because the stimulus (eating a hot fudge sundae) and the effect (eating another hot fudge sundae) are the same.

E) This is a negative feedback loop because the stimulus (eating a hot fudge sundae) and the effect (eating another hot fudge sundae) are the same.

Answer: D

35) Which of the following best defines tissue?

A) the basic structural and functional unit of an organism

B) the molecules that form the body's structure

C) a group of cells and the surrounding materials that work together to perform a particular function

D) a group of related organs with a common function

E) the membranes that cover organs

Answer: C

36) ALL of the following would be considered signs of infection EXCEPT:
   A) skin lesions of chicken pox.
   B) elevated body temperature.
   C) swollen lymph nodes.
   D) dull pain localized in the back of the neck.
   E) enlargement of the liver.

   Answer: D

37) Which of the following body systems provides protection against disease and returns proteins and plasma to the cardiovascular system?
   A) respiratory
   B) urinary
   C) endocrine
   D) lymphatic
   E) integumentary

   Answer: D

38) Assessment of body structure and function by touching body surfaces with the hands is called
   A) auscultation.
   B) percussion.
   C) palpation.
   D) autopsy.
   E) epidemiology.

   Answer: C

39) Which of the following structures is located in the mediastinum?
   A) heart
   B) lungs
   C) brain
   D) liver
   E) Both A and B are correct.

   Answer: A

40) Which of the following is TRUE regarding the skeletal system?
   A) It provides support and protection.
   B) It stores minerals.
   C) It assists in body movements.
   D) It houses cells that give rise to blood cells.
   E) All of the above are true.

   Answer: E

41) Which of the following is located in the pelvic cavity?
   A) uterus
   B) spleen
   C) gallbladder
   D) stomach
   E) Both A and C are correct.

   Answer: A

42) ALL of the following are found in the thoracic cavity EXCEPT the
   A) thymus.
   B) lungs.
   C) trachea.
   D) larynx.
   E) esophagus.

   Answer: D

43) Which of the following is considered to be the body's "internal environment" when discussing homeostasis?
   A) intracellular fluid
   B) plasma
   C) interstitial fluid
   D) hormones
   E) lymph

   Answer: C

44) A person in an anatomical position will exhibit ALL of the following EXCEPT
   A) standing erect.
   B) facing observer.
   C) feet flat on floor.
   D) arms at sides.
   E) palms against the lateral sides of the thighs.

   Answer: E

45) A sonogram is produced by
   A) the response of protons to a pulse of radio waves while they are being magnetized.
   B) comparison of an x-ray of a body organ before and after a contrast dye has been injected into a blood vessel.
   C) an x-ray beam moving in an arc around the body.
   D) high-frequency sound waves transmitted to a video monitor.
   E) computer interpretation of radioactive emissions from injected substances.

   Answer: D

46) Ginny is six years old and has grown 12 inches during the last year. Her family physician has referred her to an endocrinologist. This is because he suspects there is a problem with her
   A) blood flow to the bones.
   B) bone tissue structure.
   C) hormone balance.
   D) nerve impulse transmission.
   E) ability to absorb nutrients.

   Answer: C

47) Ginny has grown 12 inches during the last year. Her endocrinologist has ordered an MRI, possibly because he suspects a
   A) problem with the growth plates of the bones.
   B) tumor growing on a gland.
   C) problem with blood flow to the bones.
   D) malfunction of the metabolism in part of her brain.
   E) problem with her heart valves.

   Answer: B

48) Which of the following would be considered a symptom of disease?
   A) tremors in the hands
   B) excessive urine output
   C) a skin rash
   D) dizziness
   E) diarrhea

   Answer: D

49) In order to discuss the details of the metabolism of a cell, which level of structural organization would it be most helpful to understand?
   A) chemical
   B) tissue
   C) system
   D) organ
   E) organism

   Answer: A

50) When dividing the abdominopelvic cavity into quadrants, a transverse plane and a midsagittal plane are passed through the
   A) tops of the hip bones.
   B) heart.
   C) nipples.
   D) diaphragm.
   E) umbilicus.

   Answer: E

51) Place the following in correct sequence from the simplest to most complex:
      1. molecules
      2. atoms
      3. tissues
      4. cells
      5. organ
   A) 1, 2, 3, 4, 5.
   B) 2, 1, 4, 3, 5.
   C) 2, 1, 3, 4, 5.
   D) 1, 2, 4, 3, 5.

   Answer: B

52) Histology could be defined as a study of
   A) cells.
   B) tissues.
   C) chemistry of cells.
   D) gross structures of the body.

   Answer: B

53) Which statement is not true concerning characteristics of life?
   A) All body cells exhibit irritability to some extent.
   B) Each organ system is isolated from all other body systems.
   C) Growth can be an increase in size due to an increase in the number of cells.
   D) Reproduction occurs on both the cellular and organismal level.

   Answer: B

54) Homeostasis is the condition in which the body maintains
   A) the lowest possible energy usage.
   B) a relatively stable internal environment, within limits.
   C) a static state with no deviation from preset points.
   D) a changing state, within an unlimited range.

   Answer: B

55) Characteristics of life include
   A) metabolism and differentiation.
   B) responsiveness and reproduction.
   C) growth.
   D) movement.
   E) All answers are correct.

   Answer: E

56) What is the anatomical term for the chest cavity?
   A) mediastinum
   B) pneumothorax
   C) pleural cavity
   D) thoracic cavity
   E) ventral cavity

   Answer: C

57) In locating the nipple of the breast, which of the following pairs of terms would correctly apply?
   A) ventral and lateral
   B) proximal and external
   C) inferior and lateral
   D) parietal and anterior
   E) volar and caudal

   Answer: A

58) Which plane vertically divides the body through the midline into right and left portions?
   A) coronal
   B) transverse
   C) superior
   D) midsagittal
   E) frontal

   Answer: D

59) Human bodies are able to maintain a certain constancy of their internal environment. This statement
   A) refers only to the physiology of the vascular (circulatory) system.
   B) refers to homeostasis in the body.
   C) tells how positive feedback mechanisms work.
   D) refers to the direct control of cell activities by nucleic acids.
   E) is not true in very old or very young humans.

   Answer: B

60) Anabolism is the
   A) breakdown of matter.
   B) expulsion of matter.
   C) synthesis of matter.
   D) All of the answers are correct.

   Answer: C

61) _____ is the sum of all chemical processes that occur in the body.
   A) Reproduction
   B) Growth
   C) Metabolism
   D) Responsiveness

   Answer: C

62) _____ is the change that a cell undergoes from an unspecialized one to a specialized one.
   A) Metabolism
   B) Growth
   C) Movement
   D) Responsiveness

   Answer: A

63) The homeostatic responses of the body are regulated by what two systems?
   A) endocrine, nervous
   B) digestive, urinary
   C) reproductive, endocrine
   D) skeletal, muscular
   E) cardiovascular, respiratory

   Answer: A

64) A feedback system consists of three basic components: a control center, a receptor, and a (an)
   A) monitor.
   B) power pack.
   C) integrator.
   D) effector.
   E) regulator.

   Answer: D

65) If a response enhances the original stimulus, the system is classified as a _____ feedback system.
   A) neutral
   B) polarized
   C) deficit
   D) negative
   E) positive

   Answer: E

66) Physiology is primarily concerned with the
   A) growth of living things.
   B) activities of parts of the body.
   C) changing of certain cells into other kinds of cells.
   D) development of parts of the body.
   E) relation of human beings to their surroundings.

   Answer: B

67) Which of the following organ systems has the most influence on controlling and integrating body activities?
   A) respiratory
   B) digestive
   C) endocrine
   D) cardiovascular
   E) urinary

   Answer: C

68) The microscopic study of tissues is known as
   A) embryology.
   B) gross anatomy.
   C) pathology.
   D) cytology.
   E) histology.

   Answer: E

MATCHING. Choose the item in column 2 that best matches each item in column 1.

69) Column 1: brachial

Column 2: arm

Answer: arm

70) Column 1: cervical

Column 2: neck

Answer: neck

71) Column 1: otic

Column 2: ear

Foil: eye

Answer: ear

72) Column 1: crural

Column 2: lower leg

Answer: lower leg

73) Column 1: cephalic

Column 2: head

Answer: head

74) Column 1: carpal

Column 2: wrist

Foil: hand

Answer: wrist

75) Column 1: calcaneal

Column 2: heel of foot

Answer: heel of foot

76) Column 1: plantar

Column 2: sole of foot

Answer: sole of foot

77) Column 1: popliteal

Column 2: hollow behind knee

Foil: knee cap

Answer: hollow behind knee

78) Column 1: axillary

Column 2: armpit

Answer: armpit

79) Column 1: integumentary system

Column 2: provides protection from external stresses; helps regulate body temperature

Answer: provides protection from external stresses; helps regulate body temperature

80) Column 1: skeletal system

Column 2: provides support and protection; stores minerals and lipids

Answer: provides support and protection; stores minerals and lipids

81) Column 1: muscular system

Column 2: produces movement; generates heat; stabilizes body position

Answer: produces movement; generates heat; stabilizes body position

82) Column 1: nervous system

Column 2: detects, interprets, and responds to changes in the environment; regulates homeostasis

Answer: detects, interprets, and responds to changes in the environment; regulates homeostasis

83) Column 1: cardiovascular system

Column 2: transports oxygen, carbon dioxide, nutrients, and wastes to and from cells

Answer: transports oxygen, carbon dioxide, nutrients, and wastes to and from cells

84) Column 1: endocrine system

Column 2: regulates body activities via hormone secretion

Answer: regulates body activities via hormone secretion

85) Column 1: lymphatic system

Column 2: returns proteins and fluids to blood and transports fats from gastrointestinal tract to blood

Answer: returns proteins and fluids to blood and transports fats from gastrointestinal tract to blood

86) Column 1: respiratory system

Column 2: transfers oxygen and carbon dioxide between air and blood; helps regulate acid-base balance

Answer: transfers oxygen and carbon dioxide between air and blood; helps regulate acid-base balance

87) Column 1: digestive system

Column 2: breaks down food, absorbs nutrients, and eliminates wastes

Answer: breaks down food, absorbs nutrients, and eliminates wastes

88) Column 1: urinary system

Column 2: regulates volume and chemical composition of blood; helps regulate red blood cell production; eliminates wastes

Answer: regulates volume and chemical composition of blood; helps regulate red blood cell production; eliminates wastes

SHORT ANSWER. Write the word or phrase that best completes each statement or answers the question.

89) The study of functional changes associated with disease and aging is called _____.

Answer: pathophysiology

90) Listening to body sounds to evaluate the functioning of certain organs is called _____.

Answer: auscultation

91) _____ is the ability of an organism to detect and react to changes in the external or internal environment.

Answer: Responsiveness

ESSAY. Write your answer in the space provided or on a separate sheet of paper.

92) Identify and briefly define the six important life processes of the human body described in Chapter 1.

Answer: Metabolism is the sum of all chemical processes that occur in the body. Responsiveness is the ability to detect and respond to environmental changes. Movement is the motion of the body or its parts. Growth is an increase in the size and/or number of cells. Differentiation is the development of specialized from unspecialized cells. Reproduction is the formation of new cells or a new individual.

93) Mr. Barry is experiencing pain in his left hypochondriac and left lumbar regions. He has just been admitted to the emergency room following a construction accident in which he was pulled from beneath the rubble of a collapsed wall. One intern yells, "Get him to X-Ray now!" Yet another intern orders him to CT scanning. What would each intern hope to determine from the procedure ordered?

Answer: X-rays (radiography) will give good, clear indications of broken bones (such as ribs), while a CT scan would give a better indication of soft organ damage (such as a ruptured spleen).

94) Define the term homeostasis. Identify the components of a typical feedback loop, and describe the role of each.

Answer: Homeostasis is a condition of equilibrium in the body's internal environment produced by the interplay of all the body's regulatory processes. Homeostasis is regulated by feedback loops, which typically consist of a receptor, a control center, and an effector. The receptor monitors changes (stimuli) in controlled conditions and sends this information to the control center. The control center compares this input with other information from other receptors, and notifies an effector to make an appropriate change. The effector makes the appropriate response, as dictated by the control center.

95) Explain how a positive feedback loop differs from a negative feedback loop.

Answer: In a positive feedback loop, the response of the effector enhances or amplifies the original stimulus; that is, the condition is moved further away from homeostasis. In a negative feedback loop, the response of the effector is the opposite of the original stress and tends to move the controlled condition back toward homeostasis.

96) Identify and describe the locations of the major body fluid compartments. Which is most often called the body's "internal environment?"

Answer: Intracellular fluid (ICF) is the fluid within cells. Extracellular fluid (ECF) is the fluid outside of cells. Plasma is the ECF within blood vessels. Interstitial fluid is the fluid surrounding cells and is considered the internal environment.

97) Osmometer cells in the brain sense an increase in the concentration of plasma. This information is sent to the hypothalamus, which notifies the pituitary gland to release the hormone, ADH. ADH causes the kidney to save water, which lowers the concentration of the plasma.
Identify the elements of a feedback loop in this scenario. Is this a positive or a negative feedback loop? Explain your answer.

Answer: The controlled condition is plasma concentration, and increased plasma concentration is the stimulus. Osmometer cells are the receptors because they sense the increased concentration. The hypothalamus is the control center, which receives the input from the receptors and notifies the effector of the appropriate response. The pituitary gland and the kidneys act as effectors, since both are required to carry out the response. This is a negative feedback loop because the original stimulus (increased plasma concentration) is reversed.

98) Consider the following situation as a feedback loop: Tuition at a small college has been $1000 per semester for many years, and the student enrollment has remained constant at 1000 students for an equal number of years. This income exactly covers the expenses of faculty/staff salaries. The college administrators voted this year to fund a raise for the faculty and staff by raising tuition to $2000 per semester. Now only 300 students are enrolled at the college. Continue this story as a negative feedback loop.

Answer: There are many possible answers to this question, but the usual answer is to forget the raise and to lower tuition so that the students come back.

99) Consider the following situation as a feedback loop: Tuition at a small college has been $1000 per semester for many years, and the student enrollment has remained constant at 1000 students for an equal number of years. This income exactly covers the expenses of faculty/staff salaries. The college administrators voted this year to fund a raise for the faculty and staff by raising tuition to $2000 per semester. Now only 300 students are enrolled at the college. Continue this story as a positive feedback loop.

Answer: There are several correct answers to this question, but the usual answer is that the administrators raise tuition again, enrollment drops further, and the school closes.

100) One of the members of your study group is insisting that a feedback loop is a positive feedback loop because it is "doing good for the body." What is wrong, if anything, with this student's thinking?

Answer: This student is confusing the terms positive and negative with good and bad. When describing feedback loops, the terms positive and negative are used in a more quantitative way to describe whether the effects of a loop amplify (increase) or reverse (decrease) a change in a controlled condition.

CHAPTER 2   The Chemical Level of Organization

MULTIPLE CHOICE. Choose the one alternative that best completes the statement or answers the question.

1) Which of the following is TRUE regarding this situation: Solution A has a pH of 7.38, and Solution B has a pH of 7.42.
   A) Solution B is more acidic than Solution A.
   B) The pH of Solution A falls within the homeostatic pH range for extracellular body fluids, but the pH of Solution B does not.
   C) Solution A contains a higher concentration of hydrogen ions than Solution B.
   D) Solution B contains a higher concentration of hydrogen ions than Solution A.
   E) Both B and C are correct.

   Answer: C

2) An object's mass is determined by
   A) the amount of matter it contains.
   B) its weight.
   C) the type of chemical bonds present in it.
   D) its state (solid, liquid, or gas).
   E) Both B and D are correct.

   Answer: A

3) The chemical symbol for sodium is
   A) S
   B) So
   C) Sd
   D) K
   E) Na

   Answer: E

4) The four elements making up about 96% of the body's mass are represented by the symbols
   A) O, Ca, H, Na
   B) O, C, H, N
   C) O, C, He, Na
   D) O, H, K, N
   E) O, Ca, H, Ni

   Answer: B

5) Which of the following carry a negative charge?
    A) protons only
    B) neutrons only
    C) electrons only
    D) both protons and electrons
    E) both electrons and neutrons
    Answer: C

6) Peptide bonds are found in
    A) carbohydrates.
    B) lipids.
    C) proteins.
    D) inorganic compounds.
    E) any type of molecule.
    Answer: C

7) The smallest unit of matter that retains the properties and characteristics of an element is the
    A) atom.
    B) molecule.
    C) proton.
    D) nucleus.
    E) electron.
    Answer: A

8) A common buffer found in human blood plasma is
    A) glucose.
    B) bicarbonate ion.
    C) sodium ion.
    D) triglycerides.
    Answer: B

9) The speed or rate of a chemical reaction is influenced by ALL of the following EXCEPT
    A) the concentration of the reactants.
    B) the temperature.
    C) the presence of catalysts or enzymes.
    D) the presence or absence of carbon.
    Answer: D

10) The biological function of a protein is determined by its
    A) primary structure.
    B) secondary structure.
    C) tertiary structure.
    D) quaternary structure.
    E) denatured structure.

    Answer: C

11) A dalton is
    A) a special type of high-energy phosphate bond.
    B) a measure of atomic electrical charge.
    C) another name for an isotope.
    D) a unit of measurement for atomic mass.
    E) a measure of the amount of energy contained in a chemical bond.

    Answer: D

12) The function of ATP is to
    A) act as a template for production of proteins.
    B) store energy.
    C) act as a catalyst.
    D) determine the function of the cell.
    E) hold amino acids together in a protein.

    Answer: B

13) What can you tell from the molecular formula $C_6H_{12}O_6$?
    A) It easily forms free radicals.
    B) It is an organic compound.
    C) The formula represents six molecules of the substance.
    D) It contains six carbon atoms.
    E) Both B and D are correct.

    Answer: E

14) Coenzymes are
    A) organic molecules derived from vitamins
    B) two enzymes that perform the same function.
    C) metal ions.
    D) enzymes that work together.

    Answer: A

15) ALL of the following are organic molecules EXCEPT
   A) ATP.
   B) glucose.
   C) DNA.
   D) enzymes.
   E) water.
   Answer: E

16) The most abundant inorganic substance in the human body is
   A) glucose.
   B) fat.
   C) ATP.
   D) water.
   E) iron.
   Answer: D

17) Which of the following is considered to be neutral on the pH scale?
   A) urine
   B) pure water
   C) blood plasma
   D) cytoplasm
   E) interstitial fluid
   Answer: B

18) Carbohydrates are stored in the liver and muscles in the form of
   A) glucose.
   B) triglycerides.
   C) glycogen.
   D) cholesterol.
   Answer: C

19) Which of the following is a basic pH?
   A) 7
   B) 5
   C) 1
   D) 9
   F) 3
   Answer: D

20) Enzymes and antibodies are examples of
   A) carbohydrates.
   B) lipids.
   C) proteins.
   D) nucleic acids.

   Answers: C

21) In proteins, the folding of the unit on itself is referred to as its
   A) primary structure.
   B) secondary structure.
   C) tertiary structure.
   D) quaternary structure.
   E) All answers are correct.

   Answer: C

22) The most abundant chemical compound found in the human body is
   A) lipid.
   B) protein.
   C) carbohydrate.
   D) water.
   E) nucleotide.

   Answer: D

23) Isotopes of an element have
   A) the same atomic number.
   B) the same atomic mass.
   C) different numbers of protons.
   D) different numbers of electrons.
   E) the same number of neutrons.

   Answer: A

24) Amino acids
   A) contain nitrogen.
   B) are true acids.
   C) are present in carbohydrates.
   D) are present in cellulose.
   E) are also called fatty acids.

   Answer: A

25) A molecule of fat
   A) may be stored as part of a protein molecule.
   B) is converted into amino acids.
   C) is known as a protein saver.
   D) can repair or replace protoplasm.
   E) is composed of one molecule of glycerol and three molecules of fatty acids.

   Answer: E

26) An example of a polysaccharide is a
   A) histidine.
   B) lactose.
   C) glucose.
   D) lactic acid.
   E) glycogen.

   Answer: E

27) A molecule of fat
   A) may be stored as part of a protein molecule.
   B) is converted into amino acids.
   C) is known as a protein saver.
   D) can repair or replace protoplasm.
   E) is composed of one molecule of glycerol and three molecules of fatty acids.

   Answer: E

28) An example of a polysaccharide is a
   A) histidine.
   B) lactose.
   C) glucose.
   D) lactic acid.
   E) glycogen.

   Answer: E

29) The function of a catalyst is to
   A) convert strong acids and bases into weak acids and bases.
   B) store energy released during exergonic reactions.
   C) act as the chemical link between atoms in a covalent bond.
   D) lower the activation energy needed for a chemical reaction to occur.
   E) keep particles in a colloid from settling out.

   Answer: D

30) Which of the following represents accurate base-pairing in DNA molecules?
   A) adenine to adenine and guanine to guanine
   B) adenine to uracil and cytosine to guanine
   C) adenine to cytosine and guanine to thymine
   D) adenine to thymine and cytosine to guanine
   E) adenine to guanine and cytosine to thymine

   Answer: D

31) In the body, decomposition reactions are
   A) anabolic reactions.
   B) exchange reactions.
   C) catabolic reactions.
   D) usually endergonic reactions.
   E) Both C and D are correct.

   Answer: C

32) In forming a covalent bond, electrons are
   1. given up by one atom to another.
   2. shared by two atoms.
   3. taken up by one atom from another.
   A) 1
   B) 2
   C) 3
   D) 1 and 2
   E) 1, 2, and 3

   Answer: B

33) The atom of one element is distinguished from an atom of another element by the number of
   A) neutrons in the nucleus.
   B) electrons in the nucleus.
   C) protons in the nucleus.
   D) electrons orbiting the nucleus.
   E) electrons it can lose when bonding.

   Answer: C

34) The atomic number of an atom is the
   A) sum of the numbers of subatomic particles.
   B) number of electrons in the outer orbital shell.
   C) number of neutrons in the nucleus.
   D) number of protons in the nucleus.
   E) sum of the numbers of protons and neutrons only.

   Answer: D

35) Which of the following best represents the structure of a carbohydrate?

   A)   H H H

          H—C—C—C—COOH

             H H NH_3

   B)   H H H H

          H—C—C—C=C—C=O

             H H   OH

   C)   H OH H

         H—C—C—C—COOH

           OH H OH

   Answer: C

36) Solutes that are hydrophilic are those that
   A) have been reduced by addition of hydrogen atoms.
   B) dissolve easily in water because they contain polar covalent bonds.
   C) dissolve easily in water because they contain nonpolar covalent bonds.
   D) dissolve poorly because they contain polar covalent bonds.
   E) dissolve poorly because they contain nonpolar covalent bonds.

   Answer: B

37) Which one of the following nitrogenous bases will form a hydrogen bond with guanine?
   A) adenine
   B) cytosine
   C) thymine
   D) uracil

   Answer: B

38) Nucleotides contain
   A) amino acids.
   B) sugar.
   C) fatty acids.
   D) dipeptides.
   E) None of the above.

   Answer: B

39) Glycerol is the backbone molecule for
   A) disaccharides.
   B) DNA.
   C) peptides.
   D) triglycerides.
   E) ATP.

   Answer: D

40) Neutral atoms contain the same number of electrons as_____
   A) protons.
   B) neutrons.
   C) nuclei.
   D) positrons.

   Answer: A

41) The four most common elements in the human body are
   A) carbon, hydrogen, oxygen, and iron.
   B) carbon, hydrogen, oxygen, and chlorine.
   C) carbon, oxygen, nitrogen, and chlorine.
   D) carbon, hydrogen, nitrogen, and calcium.
   E) carbon, oxygen, hydrogen, and nitrogen.

   Answer: E

42) Two carbohydrate molecules are joined together by a
   A) hydrogen bond.
   B) glycosidic bond.
   C) peptide bond.
   D) ester bond.
   E) All are correct.

   Answer: B

43) The four most common ions in the human body are
    A) sodium.
    B) potassium.
    C) chloride.
    D) calcium.
    E) All answers are correct.

    Answer: E

44) Two amino acids are joined together by a
    A) hydrogen bond.
    B) glycosidic bond.
    C) peptide bond.
    D) ester bond.
    E) All are correct.

    Answer: C

45) The pH scale measures
    A) total electrolyte concentration.
    B) ATP levels.
    C) the level of enzyme activity.
    D) hydrogen ion concentration.
    E) total hydrogen content of all organic compounds in the body.

    Answer: D

46) A fatty acid with only single covalent bonds is said to be
    A) oxidized.
    B) reduced.
    C) saturated.
    D) denatured.
    E) hydrolyzed.

    Answer: C

47) A "saturated" fat is saturated with
    A) oxygen.
    B) hydrogen.
    C) carbon.
    D) ATP.
    E) nitrogen.

    Answer: B

48) The sugar found in RNA is
   A) fructose.
   B) glucose.
   C) sucrose.
   D) ribose.
   E) deoxyribose.
   Answer: D

49) The nonprotein portion of an enzyme is the
   A) apoenzyme.
   B) active site.
   C) cofactor.
   D) substrate.
   E) holoenzyme.
   Answer: C

50) ALL of the following are TRUE regarding DNA and RNA EXCEPT
   A) Both contain pentose sugars.
   B) Both contain adenine and guanine.
   C) Both contain purines and pyrimidines.
   D) Both have a double helix structure.
   E) Three types of RNA act to carry out the instructions coded in DNA.
   Answer: D

SHORT ANSWER. Write the word or phrase that best completes each statement or answers the question.

51) Substances that can speed up chemical reactions without being altered themselves are known as
   _____.
   Answer: catalysts

52) The element that "saturates" a saturated fat is _____.
   Answer: hydrogen

53) The complete hydrolysis of proteins yields _____.
   Answer: amino acids

54) The pentose sugar found in DNA is _____.
   Answer: deoxyribose

55) A molecule that gains hydrogen atoms during chemical reactions in the body is said to be _____.

Answer: reduced

56) The sum of its protons and neutrons is the _____.

Answer: mass number

57) Atoms of an element that have different numbers of neutrons are _____.

Answer: isotopes

58) The standard unit for measuring the mass of atoms and their subatomic particles is a _____.

Answer: dalton.

59) If an atom gives up or gains electrons, it becomes an _____.

Answer: ion.

60) An electrically charged atom with an unpaired electron in its outermost shell is a _____.

Answer: free radical.

61) When two atoms share one electron pair, this is a _____.

Answer: covalent bond

62) A polar covalent bond that forms between hydrogen atoms and other atoms is a _____.

Answer: hydrogen bond

63) Cohesion of water molecules creates a high _____.

Answer: surface tension

64) Reactions that release energy are called _____.

Answer: exergonic reactions

65) Chemical compounds that speed up chemical reactions by lowering the activation energy are _____.

Answer: catalysts

66) Synthetic reactions that occur in the body are referred to as_____.

Answer: anabolism

67) The compound that is the most important and most abundant inorganic compound in living organisms is _____.

Answer: water

68) Solutes that contain polar covalent bonds are _____.

Answer: hydrophilic

69) An _____ is a substance that dissociates into one or more hydrogen ions.

Answer: acid

70) A chemical that can convert a strong acid or base into a weak one is a ____.

Answer: buffer.

71) A group of monosaccharides joined together are called____.

Answer: polysaccharides

72) A three-carbon molecule that is the backbone of a triglyceride is a ____.

Answer: glycerol

73) A fatty acid that contains one or more double bonds is considered ____.

Answer: unsaturated

74) Amino acids are connected together by ____ bonds.

Answer: peptide

75) The three-dimensional shape of a polypeptide chain is its _____ structure.

Answer: tertiary

MATCHING. Choose the item in column 2 that best matches each item in column 1.

76) Made up of amino acids.   A. Monosaccharides
77) Double helix structure.   B. Proteins
78) Glucose, fructose and ribose are examples.   C. DNA
79) Starch and glycogen are examples.   D. Fatty acids
80) Hydrophobic component of a phospholipid.   E. Polysaccharides

Answers: 76) B.   77) C.   78) A.   79) E.   80) D.

81) Electrolytes that are hydrogen ion donors.   A. Bases
82) Electrolytes that are hydrogen acceptors.   B. Water
83) Acts as a solvent for ionic compounds in body fluids.   C. Acids
84) Substances that cannot be split into simpler substances.   D. Electrons
85) Found in the valence shell of an atom.   E. Elements

Answers: 81) C.   82) A.   83) B.   84) E.   85) D.

86)	Sodium.	A.	K
87)	Potassium.	B.	Ca
88)	Iron.	C.	Na
89)	Calcium.	D.	Fe
90)	Oxygen.	E.	O

Answers: 86) C.   87) K.   88) D.   89) B.   90) E.

91)	Collagen in bone.	A.	Structural
92)	Muscle cells.	B.	Regulatory
93)	Hormones.	C.	Contractile
94)	Salivary amylase.	D.	Immunological
95)	Antibodies.	E.	Catalytic

Answers: 91) A.   92) C.   93) B.   94) E.   95) D.

ESSAY. Write your answer in the space provided or on a separate sheet of paper.

96)   Describe/discuss the structural arrangement of the subatomic particles of an atom.

Answer: The positively charged protons and the uncharged neutrons are located in the nucleus of the atom. Negatively charged electrons, equal in number to the protons, orbit the nucleus in various energy levels, depending on the total number of electrons.

97)   Compare and contrast ionic vs. covalent chemical bonds.

Answer: Chemical bonding involves either donating/receiving electrons or sharing electrons between atoms. In an ionic bond, atoms are held together by the opposite electrical charges created by donating or receiving electrons. In a covalent bond, two atoms share electrons to stabilize their outer electron orbitals without loss or gain of electrons. Most organic compounds are held together by covalent bonds. Electrolytes are held together by ionic bonds.

98)   Explain the role of water in maintenance of body temperature.

Answer: Because water has such a high heat capacity, it can absorb and release large amounts of heat with only a modest change in its own temperature, thus lessening the impact of changes in environmental temperatures on the body. Also, because of the high heat of vaporization of water, the evaporation of sweat acts as an important cooling mechanism.

99)   Compare and contrast the structures of DNA and RNA.

Answer: DNA and RNA are both formed from smaller units, known as nucleotides. DNA nucleotides contain a phosphate group, deoxyribose, and a nitrogenous base – either adenine, thymine, guanine, or cytosine. RNA nucleotides contain a phosphate group, ribose, and a nitrogenous base – either adenine, uracil, guanine, or cytosine. DNA molecules are made of two strands of nucleotides joined by complementary base-pairing (A-T and C-G). RNA molecules are a single strand of nucleotides and are much shorter than DNA molecules, having been formed from a fraction of a DNA molecule.

100) What are hydrogen bonds, and what is their importance to protein structure?

Answer: Hydrogen bonds form when the hydrogen atom in a polar covalent bond develops a partial positive charge that attracts the partial negative charge of neighboring electronegative atoms. Hydrogen bonds are important for stabilizing the secondary structure of proteins.

CHAPTER 3   The Cellular Level of Organization

MULTIPLE CHOICE. Choose the one alternative that best completes the statement or answers the question.

1) The predominant lipids in the lipid bilayer of human plasma membranes are
   A) phospholipids.
   B) glycolipids.
   C) cholesterols.
   D) lipoproteins.
   E) steroids.

   Answer: A

2) The term chromatin refers to
   A) the color of certain cells.
   B) uncoiled DNA.
   C) highly coiled DNA.
   D) the fluid within the nucleus.
   E) the protein that makes up the mitotic spindle fibers.

   Answer: B

3) Organelles that contain enzymes that destroy material engulfed by phagocytes are
   A) mitochondria.
   B) lysosomes.
   C) ribosomes.
   D) nucleoli.
   E) centrioles.

   Answer: B

4) Organelles that contain enzymes for production of ATP are
   A) mitochondria.
   B) lysosomes.
   C) ribosomes.
   D) nucleoli.
   E) centrioles.

   Answer: A

5) How are phospholipid molecules arranged within the lipid bilayer of the plasma membrane?
   A) Phospholipid molecules are arranged randomly.
   B) The polar lipid tails are oriented toward the ECF and the ICF because they are hydrophobic.
   C) The polar phosphate heads are oriented toward the ECF and the ICF because they are hydrophilic.
   D) The nonpolar lipid tails are oriented toward the ECF and the ICF because they are hydrophobic.
   E) The nonpolar phosphate heads are oriented toward the ECF and the ICF because they are hydrophobic.

   Answer: C

6) The reason the lipid bilayer of a plasma membrane is asymmetric is that
   A) the phospholipids are randomly arranged.
   B) glycolipids make up the outer layer, while phospholipids make up the inner layer.
   C) cholesterol molecules line up on the inner surface of the membrane.
   D) the phospholipids in the outer layer are larger molecules than those of the inner layer.
   E) glycolipids appear only on the extracellular side of the membrane.

   Answer: E

7) Powerhouses of the cell, the most important site of ATP production is found in
   A) mitochondria.
   B) plasma membrane.
   C) Golgi apparatus.
   D) centrioles.
   E) lysosomes.

   Answer: A

8) The cell does not need to expend energy (ATP) in order to perform
   A) osmosis.
   B) pinocytosis.
   C) facilitated diffusion.
   D) active transport.
   E) Both A and C are correct.

   Answer: E

9) Carrier molecules within the plasma membrane are required in order to transport a substance across a membrane via

A) osmosis.

B) filtration.

C) facilitated diffusion.

D) simple diffusion.

E) exocytosis.

Answer: C

10) The term mediated transport refers to

A) transport of materials into the nucleus, which is in the middle of the cell.

B) passage of materials into the space in the middle of the two layers of the lipid bilayer.

C) passive transport of materials at an average rate.

D) passage of materials across a membrane with the assistance of a transporter protein.

E) passage of materials across a membrane that requires the hydrolysis of ATP.

Answer: D

11) Solutes move "down" a concentration gradient in

A) primary active transport.

B) secondary active transport.

C) osmosis.

D) facilitated diffusion.

E) phagocytosis.

Answer: D

12) The RNA responsible for bringing the amino acids to the "factory" site for protein formation is the

A) rRNA.

B) mRNA.

C) tRNA.

D) ssRNA.

Answer: C

13) Symporters are transporters that

A) move two substances in the same direction across a membrane.

B) move two substances in opposite directions across a membrane.

C) can move a substance both into and out of a cell.

D) move materials into the nucleus.

E) move materials over the surface of a cell.

Answer: A

14) Cells try to move sodium ions from the cytoplasm to the outside of the cell where the sodium concentration is 14 times higher than in the cytoplasm. This means sodium ions are moved out of the cells by

A) simple diffusion.

B) facilitated diffusion.

C) osmosis.

D) active transport.

E) exocytosis.

Answer: D

15) The genetic information is coded in DNA by

A) the regular alternation of sugar and phosphate molecules.

B) the sequence of the nucleotides.

C) the three-dimensional structure of the double helix.

D) the structure of the histones.

Answer: B.

16) The advantage of the presence of microvilli on cell membranes is

A) Such cells can move.

B) Materials can be pushed across the cell surface.

C) The membrane has a greater surface area for an increased rate of diffusion.

D) The microvilli can extend out to engulf solid particles.

E) The membrane is more permeable to polar substances.

Answer: C

17) A red blood cell placed in pure water would

A) shrink.

B) swell initially, then shrink as equilibrium is reached.

C) neither shrink nor swell.

D) swell and burst.

Answer: D

18) A function of mitosis is

A) formation of sex cells.

B) growth and cell replacement.

C) to create diversity in genetic potential.

D) All of these.

Answer: B

19) Osmosis is considered a special case of diffusion because
   A) a solute moves against its gradient.
   B) it is the net movement of solvent down its own gradient.
   C) water is moving against its own gradient.
   D) water requires a transporter protein to get through the lipid bilayer.
   E) Both C and D are correct.

   Answer: B

20) Aquaporins are
   A) the places where receptors bind to ligands.
   B) the "rivers" of water flow throughout the cytosol.
   C) sites of hydrolysis within the cytoplasm.
   D) water channels in a plasma membrane.
   E) the sugar molecules that attract a fluid layer to the cell's surface.

   Answer: D

21) The shrinking of cells when placed in a hypertonic solution is called
   A) autophagy.
   B) cytokinesis.
   C) cytolysis.
   D) denaturation.
   E) crenation.

   Answer: E

22) The end result of mitosis is
   A) two diploid cells identical to the parent cell.
   B) two haploid cells identical to the parent cell.
   C) sperm cells and egg cells.
   D) two cells with twice as much DNA as the parent cell.
   E) one cell with twice as many organelles as the parent cell.

   Answer: A

23) Facilitated diffusion is said to have a transport maximum because
   A) when all the transporters are occupied, the rate of facilitated diffusion cannot be increased further.
   B) there is a maximum size of particle that can be transported by facilitated diffusion.
   C) the concentration gradient has to be of a certain steepness for facilitated diffusion to "kick in."
   D) a cell can only make sufficient ATP up to a certain maximum rate to fuel facilitated diffusion.
   E) once electrical charges equalize, facilitated diffusion shuts down.

   Answer: A

24) The Na$^+$/K$^+$ATPase transports:
   A) both sodium and potassium ions into a cell.
   B) both sodium and potassium ions out of a cell.
   C) sodium ions into a cell and potassium ions out of a cell.
   D) sodium ions out of a cell and potassium ions into a cell.
   E) ATP across a plasma membrane.

   Answer: D

25) The following items are involved in the manufacture of proteins in a human cell. Place them in correct sequence: 1=tRNA. 2=DNA. 3= mRNA. 4=ribosomes. 5=nuclear pores.
   A) 2, 3, 5, 4, 1
   B) 2, 3, 1, 5, 4
   C) 5, 2, 3, 4, 1
   D) 3, 2, 5, 1, 4

   Answer: B

26) If the nucleotide or base sequence of the DNA strand used as a template for messenger RNA synthesis is ACGTT, then the sequence of bases in the corresponding mRNA would be
   A) TGCAA
   B) ACGTT
   C) UGCAA
   D) GUACC

   Answer: C

27) Cells that can perform phagocytosis include
   A) red blood cells.
   B) certain white blood cells.
   C) endothelial cells lining blood vessels.
   D) neurons.
   E) any cell can perform phagocytosis.

   Answer: B

28) During the process of translation, the code carried by mRNA is
   A) turned into DNA.
   B) decoded into a protein.
   C) decoded into a pentose.
   D) produced by copying the DNA template.
   E) used to manufacture RNA.

   Answer: B

29) Most microfilaments are composed of

A) actin.

B) tubulin.

C) rRNA.

D) histones.

E) cholesterol.

Answer: A

30) Cilia and flagella are made mostly of

A) microfilaments.

B) intermediate filaments.

C) microtubules.

D) rRNA.

E) phospholipids.

Answer: C

31) The process of transcription involves production of

A) mRNA from a DNA template.

B) two new DNA strands from the two original strands.

C) DNA from an mRNA template.

D) an amino acid chain from an mRNA template.

E) new amino acids.

Answer: A

32) In the maintenace of the cell resting membrane potential

A) extracellular sodium levels are high.

B) cells are more permeable to Na+ than K+.

C) the steady state involves only passive processes.

D) the inside of the cell is positive relative to its outside.

Answer: A

33) What can you tell about the following nucleotide (nitrogenous base) sequence: ADENINE-URACIL-GUANINE?

A) It could be part of DNA.

B) It could be part of RNA.

C) A complementary strand of RNA nucleotides (nitrogenous bases) would be THYMINE-ADENINE-CYTOSINE.

D) The bases are linked by peptide bonds.

E) Both A and C are correct.

Answer: B

34) The subunits of a ribosome are produced in the

A) nucleolus.

B) Golgi complex.

C) centrosome.

D) glycocalyx.

E) endoplasmic reticulum.

Answer: A

35) If this nitrogenous base sequence (CYTOSINE-CYTOSINE-ADENINE) represents the nucleotides of a codon, then

A) the anticodon would be GUANINE-GUANINE-URACIL.

B) the complementary sequence on the sense strand of DNA is the same.

C) it is part of tRNA.

D) it can code for any amino acid.

E) Both A and C are correct.

Answer: A

36) Secretory proteins and membrane molecules are synthesized mainly by the

A) mitochondria.

B) rough ER.

C) centrosome.

D) Golgi complex.

E) nucleolus.

Answer: B

37) The Golgi complex is most extensive in cells that

A) do not reproduce.

B) have many mitochondria.

C) move fluids across their surfaces.

D) secrete proteins into the ECF.

E) frequently change shape.

Answer: D

38) The major type of storage vesicles produced by the Golgi complex is the

A) centrosome.

B) nucleosome.

C) glycocalyx.

D) lysosome.

E) clathrin pit.

Answer: D

39) Recycling of worn out organelles is accomplished by autophagy, which is carried out by

   A) mitochondria.
   B) the Golgi complex.
   C) ribosomes.
   D) smooth ER.
   E) lysosomes.

   Answer: E

40) Toxic hydrogen peroxide resulting from oxidation reactions is broken down by an enzyme in peroxisomes called

   A) ATPase.
   B) kinesin.
   C) catalase.
   D) kinase.
   E) polymerase.

   Answer: C

41) An electrical gradient, or membrane potential, exists across a cell membrane because in most cells the inside surface of the membrane is

   A) more negatively charged than the outside surface.
   B) more positively charged than the outside surface.
   C) more hydrophilic than the outside surface.
   D) more hydrophobic than the outside surface.
   E) richer in lipid molecules.

   Answer: A

42) The reactions of cellular respiration occur in

   A) ribosomes.
   B) mitochondria.
   C) smooth ER.
   D) the nucleolus.
   E) Both A and B are correct.

   Answer: B

43) Solution A and Solution B are separated by a selectively permeable membrane. Solution A contains 5% sodium chloride dissolved in water. Solution B contains 10% sodium chloride dissolved in water. Which of the following statements is TRUE regarding this situation?

A) Solution A has the higher osmotic pressure because it has the greater concentration of solute.

B) Solution B has the higher osmotic pressure because it has the greater concentration of solute.

C) Solution A has the higher osmotic pressure because it has the greater concentration of solvent.

D) Solution B has the higher osmotic pressure because it has the greater concentration of solvent.

E) There is no way of knowing the relative osmotic pressures of these solutions.

Answer: B

44) Solution A contains 5% NaCl dissolved in water. Solution B contains 10% NaCl dissolved in water. Which of the following best describes the relative concentrations of these solutions?

A) Solution A is hypertonic to Solution B.

B) Solution B is hypertonic to Solution A.

C) Solutions A and B are isotonic to each other.

D) Solution B is hypotonic to Solution A.

E) Both A and D are correct.

Answer: B

45) Red blood cell membranes are not normally permeable to NaCl and maintain an intracellular concentration of NaCl of 0.9%. If these cells are placed in a solution containing 9% NaCl, what would happen?

A) Nothing because the membrane is not permeable to NaCl.

B) Water will enter the cell because the intracellular fluid has a higher osmotic pressure.

C) The cell will undergo hemolysis due to membrane damage from the 9% NaCl solution.

D) The cell will undergo crenation because the extracellular solution has a higher osmotic pressure.

E) The cell will start to generate NaCl to make up the difference in concentrations.

Answer: D

46) The function of centrioles include

A) organize the mitotic spindle in cell division.

B) provide a whip-like beating motion to move substances along cell surfaces.

C) serve as the site for ribosomal RNA synthesis.

D) Both A and B are correct.

Answer: A.

47) Lysosomes
   A) are used mainly for the cell to divide.
   B) contain acid hydrolases which are potentially dangerous to the cell.
   C) maintain a highly alkaline internal environment.
   D) are the major site of protein synthesis.

   Answer: B.

48) The membranes of lysosomes contain active transport pumps to pump hydrogen ions into the lysosomes because
   A) hydrogen ions need to be destroyed.
   B) the hydrogen ions will raise the pH inside the lysosome so that the lysosomal enzymes can work more efficiently.
   C) the hydrogen ions will lower the pH inside the lysosome so that the lysosomal enzymes can work more efficiently.
   D) they need to pump sodium ions out by secondary active transport.
   E) the lysosomal enzymes need to be alternately reduced and oxidized to function.

   Answer: C

49) Which cell organelle is the site of fatty acid, phospholipid, and steroid synthesis?
   A) Golgi complex.
   B) lysosome.
   C) mitochondria.
   D) rough endoplasmic reticulum.
   E) smooth endoplasmic reticulum.

   Answer: E

50) Which of the following lists the phases of the cell cycle in the correct sequence?
   A) S phase, G1 phase, G2 phase, cytokinesis, S phase.
   B) mitosis, cytokinesis, G1 phase, G2 phase, S phase.
   C) cytokinesis, G1 phase, S phase, G2 phase, mitosis.
   D) G1 phase, G2 phase, mitosis, S phase, cytokinesis.
   E) G1 phase, S phase, G2 phase, mitosis, cytokinesis.

   Answer: E

MATCHING. Choose the item in column 2 that best matches each item in column 1.

51) _____synthesizes phospholipids, fats, and steroids.

52) _____sites of protein synthesis.

53) _____responsible for the modification, sorting and packaging of secreted products.

54) _____vesicular structures that contain enzymes such as catalase, that oxidize toxic substances.

55) _____site of mRNA synthesis.

A. Smooth ER

B. Ribosomes

C. Golgi complex

D. Peroxisomes

E. Nucleus

Answers: 51) A.   52) B.   53) C.   54) D.   55) E.

56) _____contain many types of enzymes to digest.

57) _____function to move an entire cell.

58) _____organizing center for mitotic spindle.

59) _____cellular respiration.

60) _____controls what enters and leaves a cell.

A. Lysosome

B. Flagella

C. Centrosome

D. Mitochondria

E. Plasma membrane

Answers: 56) A.   57) B.   58) C.   59) D.   60) E.

61) _____attached to an amino acid.

62) _____contained in a chromosome.

63) _____protein attached to a ribosome.

64) _____catalyzes transcription of DNA.

65) _____directs the synthesis of a protein.

A. RNA plymerase

B. mRNA

C. tRNA

D. rRNA

E. DNA

Answers: 61) C.   62) E.   63) D.   64) A.   65) B.

66) _____mitotic spindle forms.

67) _____centromeres split.

68) _____chromatid pairs align at center of cell.

69) _____cytokinesis.

70) _____DNA replication.

A. Metaphase

B. Prophase

C. Telophase

D. Anaphase

E. Interphase

Answers: 66) B.   67) D.   68) A.   69) C.   70) E.

SHORT ANSWER. Write the word or phrase that best completes each statement or answers the question.

71) If Solution A has proportionally more solutes and less water than Solution B, then Solution A is considered to be _____ to Solution B.

Answer: hypertonic

72) A group of nucleotides on a DNA molecule whose purpose is to serve as the "directions" for manufacturing a specific protein is a _____.

Answer: gene

73) The distribution of two sets of chromosomes into two separate and equal nuclei is known as _____.

Answer: mitosis

74) The cytosol and the organelles are the two components of the _____.

Answer: cytoplasm

75) Because phospholipids have both polar and nonpolar parts, they are said to be _____.

Answer: amphipathic

76) The part of a phospholipid molecule that is hydrophilic is the _____.

Answer: head (polar end or phosphate-containing end)

77) Chromosome number does not double with each generation of cell division because of a special nuclear division called _____.

Answer: meiosis

78) Membrane proteins that extend across the entire lipid bilayer are known as _____ proteins.

Answer: integral

79) The sugar coat that provides a means of cellular recognition is known as the _____.

Answer: glycocalyx

80) The property of membranes that allows some substances to pass more readily than others is called _____.

Answer: selective permeability

81) The fluid mosaic model describes the structure of the _____.

Answer: plasma membrane

82) Transmembrane proteins that function as water channels are called _____.

Answer: aquaporins

83) The part of a phospholipid molecule that lines up facing away from the intracellular and the extracellular fluids is the _____.

Answer: tail (hydrophobic, nonpolar, lipid part)

84) In facilitated diffusion, once all the transporters are occupied, it is said that the _____ has been reached.

Answer: transport maximum

85) The peripheral protein lining pits involved in the binding phase of receptor-mediated endocytosis is called _____.

Answer: clathrin

86) Projections of the plasma membrane and cytoplasm of a phagocytic cell toward its target are called _____.

Answer: pseudopods

87) The cytoskeleton includes microfilaments, intermediate filaments, and _____.

Answer: microtubules

88) There is no net movement of water molecules across a membrane separating solutions that are _____ to each other.

Answer: isotonic

89) The two subunits of ribosomes are produced in the _____.

Answer: nucleolus

90) A type of passive transport across a plasma membrane that requires special transporters is _____.

Answer: facilitated diffusion

91) Rough ER is called "rough" because its outer surface is studded with _____.

Answer: ribosomes

92) Export of substances from the cell in which vesicles fuse with the plasma membrane and release their contents into the extracellular fluid is known as _____.

Answer: exocytosis

93) The bulging edges of the membranous sacs forming the Golgi complex are called _____.

Answer: cisterns

94) The process by which lysosomes destroy the cells they are part of is called _____.

Answer: autolysis

95) The proteins around which DNA wraps in a chromatin fiber are called _____.

Answer: histones

ESSAY. Write your answer in the space provided or on a separate sheet of paper.

96) Discuss the fluid mosaic model of plasma membrane structure.

Answer: Proteins "float" among phospholipids; phospholipid bilayer, with hydrophobic ends facing each other and hydrophilic ends facing either ECF or ICF; phospholipids can move sideways within layer; glycolipids (cellular identity markers) face ECF; cholesterol molecules (for stability) among phospholipids of both layers of the bilayer; integral proteins (channels, transporters, etc.) extend across phospholipid bilayer; peripheral proteins (enzymes, cytoskeleton anchors, etc.) loosely attached to either the inner or outer surface of the membrane.

97) Name and describe the four phases of mitosis in sequence.

Answer: 1. Prophase – chromatin fibers shorten and coil into chromosomes; nucleoli and nuclear envelope disappear; centrosomes with centrioles move to opposite poles of cell; mitotic spindle appears.

2. Metaphase – centromeres of chromatid pairs line up on metaphase plate of cell.

3. Anaphase – centromeres divide; identical sets of chromosomes move to opposite poles of cell.

4. Telophase – nuclear envelope reappears to enclose chromosomes; chromosomes revert to chromatin; nucleoli reappear; mitotic spindle disappears.

98) Name the three types of RNA and describe the role of each in protein synthesis.

Answer: 1. Messenger RNA – produced from sense strand of DNA via transcription; carries code for making a particular protein from the nucleus to the ribosome

2. Ribosomal RNA – makes up the ribosome; moves along the mRNA strand to "read" directions for making protein.

3. Transfer RNA – transports specific amino acids from the cytoplasm to the growing peptide chain; places amino acids in proper sequence by matching its anticodon with an appropriate codon on the mRNA strand.

99) Compare and contrast primary and secondary active transport.

Answer: Both are mediated, energy-requiring processes moving substances against their concentration gradients. Energy obtained from hydrolysis of ATP drives primary active transport, while energy stored in an ionic concentration gradient drives secondary active transport.

100) Compare and contrast simple diffusion and osmosis.

Answer: Both are passive types of movements. Simple diffusion involves the net movement of solute molecules from an area of higher solute concentration to an area of lower solute concentration. Osmosis involves the movement of water (solvent) across a semipermeable membrane from an area of higher water concentration to an area of lower water concentration. Both may occur through the lipid bilayer or through special channels.

MULTIPLE CHOICE. Choose the one alternative that best completes the statement or answers the question.

1) Microvilli and goblet cells are typical modifications of
   A) skeletal muscle tissue.
   B) simple columnar epithelium.
   C) osseous tissue.
   D) hyaline cartilage.
   E) nervous tissue.

   Answer: B

2) The tissue that forms glands is
   A) epithelial tissue.
   B) connective tissue.
   C) nervous tissue.
   D) muscle tissue.
   E) mesenchyme.

   Answer: A

3) Fibroblasts are the typical cells of
   A) simple squamous epithelium.
   B) dense connective tissue.
   C) cardiac muscle tissue.
   D) skeletal muscle tissue.
   E) nervous tissue.

   Answer: B

4) The type of cell junction that prevents the contents of the stomach or urinary bladder from leaking into surrounding tissues is the
   A) adherens junction.
   B) gap junction.
   C) hemidesmosome.
   D) desmosome.
   E) tight junction.

   Answer: E

5) A dense layer of proteins, called plaque, forms
   A) gap junctions.
   B) the glycocalyx.
   C) adherens junctions.
   D) microvilli.
   E) the striations of skeletal muscle.

   Answer: C

6) Intermediate filaments are important components of
   A) adherens junctions.
   B) gap junctions.
   C) tight junctions.
   D) desmosomes.
   E) Both B and D are correct.

   Answer: D

7) Connexons are
   A) the typical cells of connective tissue.
   B) fibers that provide nutrient waste and gas exchange for epithelial cells.
   C) proteins found in muscle fibers.
   D) the route by which osteocytes in osseous tissue receive nutrients.
   E) proteins in gap junctions that form the channels that provide a means of ion exchange between cells at gap junctions.

   Answer: E

8) Which of the following is typical of an endocrine gland?
   A) It releases hormones.
   B) It is made of epithelial tissue.
   C) It releases its products into ducts that empty onto epithelial surfaces.
   D) Both A and B are correct.
   E) Both B and C are correct.

   Answer: D

9) Ions and small molecules can travel between cells via
   A) tight junctions.
   B) adherens junctions.
   C) gap junctions.
   D) desmosomes.
   E) Both A and C are correct.

   Answer: C

10) Which of the following tissues is avascular?
   A) adipose tissue
   B) areolar connective tissue
   C) cardiac muscle tissue
   D) skeletal muscle tissue
   E) stratified squamous epithelium

   Answer: E

11) ALL of the following are TRUE for epithelial tissues EXCEPT
   A) they attach to connective tissues via a basement membrane.
   B) they have a rich blood supply.
   C) they have a high rate of cell division.
   D) they have a nerve supply.
   E) there is little extracellular space between adjacent plasma membranes.

   Answer: B

12) Which of the following tissues provides the greatest protection from mechanical injury?
   A) stratified squamous epithelium
   B) dense regular connective tissue
   C) smooth muscle
   D) simple cuboidal epithelium
   E) elastic cartilage

   Answer: A

13) Endothelium forms
   A) endocrine glands.
   B) exocrine glands.
   C) the lining of the heart and blood vessels.
   D) the inner layer of the skin.
   E) Both A and B are correct.

   Answer: C

14) The function of the basement membrane is to
    A) provide a blood supply to epithelial tissue.
    B) hold cartilage onto bone.
    C) house the reproducing cells of stratified squamous epithelium.
    D) anchor epithelial tissues onto underlying connective tissue.
    E) secrete matrix.

    Answer: D

15) Gap junctions are seen in tissues which
    A) fluid leakage between cells is a particular risk.
    B) cells may be damaged by friction or stretching.
    C) move from one location to another.
    D) electrical or chemical signals must pass between cells.
    E) adjacent cells are many millimeters apart.

    Answer: D

16) Secretion and absorption are important functions of
    A) adipose tissue.
    B) smooth muscle.
    C) stratified squamous epithelium.
    D) simple cuboidal epithelium.
    E) Both C and D are correct.

    Answer: D

17) Microvilli and goblet cells are characteristic of
    A) areolar connective tissue.
    B) simple squamous epithelium.
    C) simple columnar epithelium.
    D) skeletal muscle tissue.
    E) All of these could have microvilli and goblet cells.

    Answer: C

18) Cilia are commonly seen on cells in tissues that
    A) line the blood vessels.
    B) line the respiratory tract.
    C) form the skin.
    D) move from one place to another.
    E) All of these are correct.

    Answer: B

19) Keratin is seen in tissues that
    A) resist friction.
    B) move throughout the body.
    C) engulf bacteria.
    D) are important for diffusion and absorption.
    E) transmit nerve impulses.

Answer: A

20) The type of epithelium seen in the urinary bladder is
    A) simple squamous epithelium.
    B) simple cuboidal epithelium.
    C) pseudostratified columnar epithelium.
    D) stratified squamous epithelium.
    E) transitional epithelium.

Answer: E

21) Mesothelium is seen in
    A) kidney tubules.
    B) serous membranes.
    C) the urinary bladder.
    D) the lining of the heart and blood vessels.
    E) skin.

Answer: B

22) Most of the upper respiratory tract is lined with
    A) hyaline cartilage.
    B) keratinized stratified squamous epithelium.
    C) ciliated pseudostratified columnar epithelium.
    D) transitional epithelium.
    E) skeletal muscle.

Answer: C

23) Which of the following is considered a unicellular exocrine gland?
    A) mast cell
    B) plasma cell
    C) fibroblast
    D) adipocyte
    E) goblet cell

Answer: E

24) The surface area for diffusion across the membranes of cells in simple columnar epithelium is increased by the presence of
   A) goblet cells.
   B) microvilli.
   C) cilia.
   D) desmosomes.
   E) All of the above increase the surface area.

   Answer: B

25) An exocrine gland in which a cell filled with a secretory product dies and becomes the secretory product is called a(n)
   A) acinar gland.
   B) apocrine gland.
   C) holocrine gland.
   D) merocrine gland.
   E) simple gland.

   Answer: C

26) One might expect to see microvilli on epithelial tissues whose principal function is
   A) protection.
   B) movement.
   C) absorption.
   D) mineral storage.
   E) transmission of electrical impulses.

   Answer: C

27) Stratified squamous epithelium can be made waterproof and friction resistant by intracellular deposits of
   A) collagen
   B) hyaluronidase.
   C) chondroitin sulfate.
   D) mucus.
   E) keratin.

   Answer: E

28) Most exocrine glands in the human body are classified as
   A) acinar glands.
   B) apocrine glands.
   C) holocrine glands.
   D) merocrine glands.
   E) endocrine glands.

   Answer: D

29) Most exocrine glands in the human body release their products by
   A) pinching them off into the apical surface of the cell.
   B) active transport through specific ion channels.
   C) death and subsequent rupture of the secretory cell.
   D) simple diffusion.
   E) exocytosis.

   Answer: E

30) The matrix of a connective tissue consists of
   A) ground substance and fibers.
   B) all of the functioning cells of a tissue.
   C) microfilaments and microtubules.
   D) an embryonic form of the connective tissue.
   E) mesoderm and endoderm.

   Answer: A

31) The most abundant tissue in the body is
   A) epithelium tissue.
   B) connective tissue.
   C) muscle tissue.
   D) nervous tissue.
   E) All tissues are equally abundant when the body is in homeostasis.

   Answer: B

32) The basic tissue type whose function is particularly related to the form of its matrix (solid, semisolid, or liquid) is
   A) epithelium tissue.
   B) connective tissue.
   C) muscle tissue.
   D) nervous tissue.
   E) The structure of the matrix is unrelated to the functioning of tissues.

   Answer: B

33) The suffix -blast in a cell name indicates a(n)
   A) cell that has ruptured.
   B) mature cell with reduced capacity for cell division.
   C) cell that is part of an exocrine gland.
   D) immature cell that can still divide.
   E) cell that is part of the stroma of an organ.

   Answer: D

34) The embryonic connective tissue from which all other connective tissues arise is
   A) areolar connective tissue.
   B) mucous connective tissue.
   C) hyaline cartilage.
   D) neuroglia.
   E) mesenchyme.

   Answer: E

35) The diffusion of injected drugs can be enhanced by the action of hyaluronidase because it
   A) increases the rate of ATP synthesis.
   B) raises the local temperature by causing inflammation, so molecules move faster.
   C) lowers the viscosity of the matrix of areolar connective tissue.
   D) forms new, tighter junctions between the cells of connective tissue.
   E) increases the rate of matrix secretion by fibroblasts.

   Answer: C

36) The principal adhesion protein of connective tissue is
   A) fibrillin.
   B) chondroitin sulfate.
   C) collagen.
   D) fibronectin.
   E) hyaluronic acid.

   Answer: D

37) ALL of the following are functions of adipose tissue EXCEPT
   A) support.
   B) protection.
   C) formation of certain glands.
   D) insulation.
   E) energy reserve.

   Answer: C

38) Nonstriated and involuntary are terms used to describe
   A) skeletal muscle tissue.
   B) smooth muscle tissue.
   C) cardiac muscle tissue.
   D) hyaline cartilage.
   E) Both B and C are correct.

   Answer: B

39) Nutrient and waste exchange between osteocytes and blood vessels in the central canals of osteons occurs via structures called
   A) lacunae.
   B) lamellae.
   C) canaliculi.
   D) trabeculae.
   E) connexons.

   Answer: C

40) The space between the parietal and visceral layers of a membrane, such as the pericardium, is normally filled with
   A) air.
   B) blood.
   C) serous fluid.
   D) synovial fluid.
   E) adipose tissue.

   Answer: C

41) Both desmosomes and gap junctions are found in the intercalated discs connecting cells of
   A) adipose tissue.
   B) osseous tissue.
   C) skeletal muscle tissue.
   D) cardiac muscle tissue.
   E) simple squamous epithelium.

   Answer: D

42) If fibrosis occurs during tissue repair, then
   A) a perfect reconstruction of the injured tissue occurs.
   B) rapid replication of parenchymal cells has occurred.
   C) the function of the repaired tissue is impaired.
   D) only connective tissue was involved in the injury.
   E) Both B and C are correct.

   Answer: C

43) The scab over a wound to the skin is formed primarily by
   A) keratin in the epidermis.
   B) calcium salts.
   C) serous fluid as it evaporates.
   D) fibrin in blood clots.
   E) collagen in torn connective tissue.

   Answer: D

44) Because of its structure, bone is usually classified as a form of
   A) connective tissue.
   B) epithelial tissue.
   C) nonliving tissue.
   D) muscle tissue.

   Answer: A

45) Cardiac muscle tissue
   A) contains many nuclei in each cell.
   B) is both striated and branched.
   C) lacks striations.
   D) is under voluntary control.
   E) is similar to smooth muscle.

   Answer: B

46) A structure associated with a Haversian canal is a
   A) bursa.
   B) tendon.
   C) ligament.
   D) lamella.
   E) joint.

   Answer: D

47) A bone-forming cell is a(n)
   A) osteoclast.
   B) Haversian canal.
   C) lamellae.
   D) osteoblast.
   E) osteocyte.

   Answer: D

48) Striated (skeletal) muscle tissue
   A) is composed of long, spindle-like cells, each containing a single nucleus.
   B) has the ability to contract when stimulated.
   C) is present in the walls of arteries.
   D) has the ability to contract rhythmically by itself.

   Answer: B

49) Chondroblasts
   A) are mature cartilage cells located in spaces called lacunae.
   B) are within the cartilage divide and secrete new matrix.
   C) are located deep to the perichondrium divide and secrete new matrix on the external surface of the cartilage.
   D) Both B and C are correct.

   Answer: D

50) Connective tissue of the nervous tissue are the
   A) neurons.
   B) dendrites.
   C) neuroglia.
   D) axons.
   E) All are correct.

   Answer: C

MATCHING. Choose the item in column 2 that best matches each item in column 1.

_____51) Produce mucus.

_____52) Cells described as being striated and voluntary.

_____53) Develop from white blood cells called B lymphocytes.

_____54) Product histamine.

_____55) May contain keratin.

A. Stratified squamous epithelial cells
B. Goblet cells
C. Mast cells
D. Skeletal muscle cells
E. Plasma cells

Answers: 51) B.   52) D.   53) E.   54) C.   55) A.

_____56) Develop from white blood cells called monocytes.

_____57) Bone cells.

_____58) Fat storage cells.

_____59) Attached to each other via intercalated discs.

_____60) Secrete collagen.

A. Fibroblasts
B. Macrophages
C. Osteocytes
D. Adipocytes
E. Cardiac muscle cells

Answers: 56) B.   57) C.   58) D.   59) E.   60) A.

_____61) Found only in the wall of the heart.

_____62) Most widely distributed tissue in the body; located under skin and around all organs.

_____63) Embryonic connective tissue from which all other connective tissues are derived.

_____64) Very tough connective tissue containing many parallel collagen fibers.

_____65) Covers ends of bones at joints; forms much of embryonic skeleton.

A. Hyaline cartilage
B. Dense regular connective tissue
C. Mesothelium
D. Fibrocartilage
E. Cardiac muscle tissue

Answers: 61) E.   62) D.   63) C.   64) B.   65) A.

_____66) An especially smooth type of simple squamous epithelium that lines heart and blood vessels.

_____67) Multiple layers of flat cells designed for protection.

_____68) Connective tissue specialized for fat storage.

_____69) Cells may have microvilli and/or cilia.

_____70) Found in serous membranes and produces serous fluid.

A. Endothelium
B. Simple columnar epithelium
C. Simple squamous epithelium
D. Adipose tissue
E. Stratified squamous epithelium

Answers: 66) A.   67) E.   68) D.   69) B.   70) C.

SHORT ANSWER. Write the word or phrase that best completes each statement or answers the question.

71) The primary function of stratified squamous epithelium is _____.

Answer: protection

72) Simple squamous epithelium that lines the heart, blood vessels, and lymphatic vessels and forms the walls of the capillaries is _____.

Answer: endothelium

73) _____ glands secrete their products into ducts.

Answer: Exocrine

74) Antibodies are produced by connective tissue cells known as _____.

Answer: plasma cells

75) The principal adhesion protein of connective tissue is _____.

Answer: fibronectin

76) The stroma of soft organs is formed by _____ fibers.

Answer: reticular

77) The cells of mature cartilage are called _____.

Answer: chondrocytes

78) The connective tissue from which all other connective tissues eventually arise is _____.

Answer: mesenchyme

79) Spaces in the matrix of cartilage or bone in which cells are located are called _____.

Answer: lacunae

80) The type of cartilage that always lacks a perichondrium is _____.

Answer: fibrocartilage

81) The type of membrane that lines a body cavity that does not open to the exterior is a _____.

Answer: serous membrane

82) The connections between cardiac muscle cells are called _____.

Answer: intercalated discs

83) The type of cartilage growth in which new matrix is added on to the surface of existing cartilage is called _____ growth.

Answer: appositional

84) Columns of bone-forming spongy bone are called _____.

Answer: trabeculae

85) The surface area of the apical surfaces of epithelial cell membranes is increased by the presence of _____.

Answer: microvilli

86) Goblet cells produce _____.

Answer: mucus

87) The liquid matrix of blood tissue is called _____.

Answer: plasma

88) Membranes lining joints are called _____ membranes.

Answer: synovial

89) The type of cell in areolar connective tissue that produces histamine is the _____.

Answer: mast cell

90) The portion of a serous membrane attached to the cavity wall is called the _____ layer.

Answer: parietal

91) The part of a neuron that contains the nucleus is the _____.

Answer: cell body

92) Cells in nervous tissue that do not generate or conduct nerve impulses are called _____.

Answer: neuroglia

93) Cells that constitute a tissue's or an organ's functioning part are called the _____.

Answer: parenchyma

94) The immature cells of each major type of connective tissue have names that end in the suffix _____.

Answer: -blast

95) The three types of fibers seen in connective tissue are _____ fibers, _____ fibers, and _____ fibers.

Answer: collagen, elastic, reticular

ESSAY. Write your answer in the space provided or on a separate sheet of paper.

96) Name the four principal types of human tissues, and briefly describe the function of each.

Answer: 1. Epithelial tissue – covers body surfaces; lines hollow organs, body cavities, and ducts; forms glands.

2. Connective tissue – protects and supports the body and its organs; binds organs together; stores energy reserves as fat; provides immunity.

3. Muscle tissue – responsible for movement and generation of force.

4. Nervous tissue – initiates and transmits action potentials (nerve impulses) that help coordinate body activities.

97) Name and describe the structures of the various types of cell junctions.

Answer: 1. Tight junctions – outer surfaces of adjacent plasma membranes fused by weblike strip of proteins.

2. Adherens junctions – made of dense layer of proteins (plaque) on inside of plasma membrane; microfilaments extend from plaque into cytoplasm; transmembrane glycoproteins anchored in plaque join adjacent cells.

3. Desmosomes – made of plaque; link adjacent cell membranes via transmembrane glycoproteins; intermediate filaments extend across cytoplasm from desmosome to desmosome in same cell.

4. Hemidesmosomes – structurally like half a desmosome, but proteins link different tissue types.

5. Gap junctions – fluid-filled gap between cells bridged by transmembrane channels (connexons).

98) Name and briefly describe the functions of the various types of cell junctions.

Answer: 1. Tight junctions – prevent passage of substances between cells.

2. Adherens junctions – help epithelial surfaces resist separation.

3. Desmosomes – attach cells to each other; structure provides tissue with greater stability.

4. Hemidesmosomes – anchor one kind of tissue to another.

5. Gap junctions – allow communication between cells.

99) Describe the basic structural characteristics of epithelial tissue.

Answer: Epithelial cells are arranged in continuous sheets, either single or multiple. Cells are packed very close together and have numerous cell junctions. There is very little extracellular material. Apical surfaces of cells are exposed to a body cavity, the lumen of a tube or organ, or the exterior of the body. The basal surfaces are attached to connective tissue via a basement membrane. Epithelial tissues are avascular but do have a nerve supply.

100) Identify and state the functions of the types of cells commonly seen in areolar connective tissue.

Answer: 1. Fibroblasts – secrete fibers and ground substance of matrix.

2. Macrophages – provide protection by engulfing bacteria and cellular debris via phagocytosis.

3. Plasma cells – provide protection via secretion of antibodies.

4. Mast cells – produce histamine, which promotes inflammation.

5. Adipocytes – specialized to store triglycerides.

6. White blood cells – provide protection during infections and allergic responses.

CHAPTER 5   The Integumentary System

MULTIPLE CHOICE. Choose the one alternative that best completes the statement or answers the question.

1)   Keratinocytes are the predominant cells in the
   A)   epidermis.
   B)   papillary region of the dermis.
   C)   reticular region of the dermis.
   D)   subcutaneous layer.
   E)   All of the above are correct.

   Answer: A

2)   Nourishment to cells in the epidermis is provided by
   A)   blood vessels running through the stratum basale.
   B)   keratinocytes.
   C)   blood vessels in the dermal papillae.
   D)   bacteria that live in sebaceous glands.
   E)   Both A and C are correct.

   Answer: C

3)   Absorption of damaging light rays is the primary function of
   A)   keratin.
   B)   sebum.
   C)   cerumen.
   D)   melanin.
   E)   keratohyalin.

   Answer: D

4)   The layer of the skin from which new epidermal cells are derived is the
   A)   stratum corneum.
   B)   stratum basale.
   C)   stratum lucidum.
   D)   dermis.
   E)   reticular layer.

   Answer: B

5) The function of keratin is to
    A) make bone hard.
    B) make skin tough and waterproof.
    C) protect skin from ultraviolet light.
    D) provide added pigment to the skin of Asian races.
    E) provide nourishment to the epidermal cells.

    Answer: B

6) The reproducing cells of the epidermis are found in the
    A) stratum basale.
    B) stratum spinosum.
    C) stratum lucidum.
    D) stratum corneum.
    E) All of these layers contain reproducing cells.

    Answer: A

7) The stratum basale contains
    A) stem cells of keratinocytes.
    B) many blood vessels.
    C) eccrine sweat glands.
    D) hair follicles.
    E) Both A and B are correct.

    Answer: A

8) Which of the following is most superficial?
    A) stratum basale
    B) papillary region of the dermis
    C) hypodermis
    D) stratum granulosum
    E) stratum corneum

    Answer: E

9) The epidermis is made up of
    A) dense irregular connective tissue.
    B) stratified squamous epithelium.
    C) areolar connective tissue.
    D) smooth muscle.
    E) All of the above are correct.

    Answer: B

10) "Goosebumps" occur due to
    A) over-stimulation of secretion from sudoriferous glands.
    B) over-stimulation of secretion from sebaceous glands.
    C) separation of the epidermis from the dermis.
    D) vasodilation of blood vessels in the skin.
    E) the action of arrector pili muscles as they raise hairs to an upright position.

    Answer: E

11) The outermost layer of the epidermis is the
    A) stratum lucidum.
    B) reticular layer.
    C) stratum corneum.
    D) superficial fascia.
    E) stratum basale.

    Answer: C

12) Sweat is produced by
    A) keratinocytes.
    B) melanocytes.
    C) ceruminous glands.
    D) sudoriferous glands.
    E) sebaceous glands.

    Answer: D

13) Which of the following is present in thick skin but not in thin skin?
    A) stratum germinativum
    B) stratum lucidum
    C) stratum corneum
    D) stratum granulosum
    E) dermal papillae

    Answer: B

14) The function of melanin is to
    A) make skin tough and waterproof.
    B) connect the epidermis to the dermis.
    C) provide flexibility to skin.
    D) provide nutrients to dying epidermal cells.
    E) protect skin from ultraviolet light.

    Answer: E

15) Hair and nails are modifications of the
    A) melanocytes.
    B) hypodermis.
    C) sudoriferous glands.
    D) epidermis.
    E) dermis.

    Answer: D

16) The stratum corneum is
    A) the innermost layer of the epidermis.
    B) highly vascular.
    C) made up of dead cells.
    D) seen only in the palms and soles.
    E) the layer in which keratin begins to form.

    Answer: C

17) Melanocytes
    A) are spidery-shaped cells in contact with cells in the stratum basale.
    B) forms structures called melanosomes.
    C) produces a substance incorporated by other cells.
    D) All answers are correct.

    Answer: D

18) Which of the following statements is TRUE regarding the epidermis?
    A) It is keratinized.
    B) Blood vessels travel from the dermis to the outer layers through special channels.
    C) All of the cells in the epidermis reproduce rapidly.
    D) It is made mostly of areolar connective tissue.
    E) Both A and C are correct.

    Answer: A

19) The "ABCD" signs are used to assess
    A) the seriousness of decubitus ulcers.
    B) whether sufficient oxygen is being transported by blood.
    C) the percentage of surface area lost to a burn.
    D) a person's total risk of developing skin cancer.
    E) a skin lesion suspected of being a malignant melanoma.

    Answer: E

20) The average length of time for a cell to be produced by the stratum basale, rise to the surface, become keratinized, and slough off is about how long?
   A) 24 minus 48 hours
   B) two weeks
   C) one month
   D) one year
   E) Once cells are keratinized, they never slough off.

   Answer: C

21) A skin condition in which abnormal keratin is produced and keratinocytes are shed prematurely is
   A) psoriasis.
   B) malignant melanoma.
   C) albinism.
   D) alopecia.
   E) impetigo.

   Answer: A

22) Corpuscles of touch (Meissner's corpuscles) are located in the
   A) stratum basale.
   B) stratum corneum.
   C) apocrine sweat glands.
   D) dermal papillae.
   E) hair follicles.

   Answer: D

23) Just beneath the stratum basale of the epidermis is the
   A) stratum corneum of the epidermis.
   B) hypodermis.
   C) reticular layer of the dermis.
   D) papillary regions of the dermis.
   E) skeletal muscle.

   Answer: D

24) Nutrients reach the surface of the skin (epidermis) through the process of
   A) absorbing material applied to the surface layer of the skin.
   B) utilizing the products of merocrine glands to nourish the epidermis.
   C) the outer layer of the skin does not require nutrients because the external layer of cells is not living.
   D) diffusing through the tissue fluid from blood vessels in the dermis.

   Answer: D

25) The papillary region of the dermis consists mostly of
   A) areolar connective tissue.
   B) adipose tissue.
   C) smooth muscle.
   D) stratified squamous epithelium.
   E) dense irregular connective tissue.

   Answer: A

26) The reticular layer of the dermis consists mostly of
   A) areolar connective tissue.
   B) adipose tissue.
   C) smooth muscle.
   D) stratified squamous epithelium.
   E) dense irregular connective tissue.

   Answer: E

27) Striae are
   A) free nerve endings sensing touch.
   B) the epidermal ridges that form fingerprints.
   C) stretch marks resulting from tears in the dermis.
   D) intermediate filaments connecting desmosomes.
   E) areas of fat storage.

   Answer: C

28) Fat storage is an important function of the
   A) epidermis.
   B) papillary region of the dermis.
   C) reticular region of the dermis.
   D) subcutaneous layer.
   E) All of the above except the epidermis.

   Answer: D

29) Tyrosinase is required for the production of
   A) keratin.
   B) melanin.
   C) cerumen.
   D) sebum.
   E) apocrine sweat.

   Answer: B

30) Synthesis of vitamin D begins with the activation of a precursor molecule in the skin by
   A) melanin.
   B) keratin.
   C) sebum.
   D) UV light.
   E) temperatures above 60 degrees F in the external environment.

   Answer: D

31) Enzymatic activity within melanosomes is increased by
   A) apoptosis of epidermal cells.
   B) environmental temperatures above normal body temperature.
   C) increased activity of sweat glands.
   D) tension on desmosomes.
   E) exposure to UV light.

   Answer: E

32) Albinism results from
   A) liver disease.
   B) low oxygen levels in the blood.
   C) lack of the enzyme tyrosinase.
   D) too little exposure to sunlight.
   E) viral infection.

   Answer: C

33) The epidermis consists of five layers of cells, each layer with a distinct role to play in the health, well-being and functioning of the skin. Which of the following layers is responsible for cell division and replacement?
   A) stratum corneum
   B) stratum granulosum
   C) stratum germinativum
   D) stratum lucidum

   Answer: C

34) Which of the following statements best describes what fingernails actually are?
   A) A modification of the epidermis.
   B) Are identical to hair but contain ten times as much keratin.
   C) Are extensions of the carpal bones.
   D) Have nothing to do with skin.

   Answer: A

35) Cyanosis is indicative of
   A) lack of tyrosinase.
   B) liver disease.
   C) inflammation.
   D) insufficient oxygen in blood.
   E) patchy loss of melanocytes.

   Answer: D

36) The blood vessels nourishing a hair follicle are located in the
   A) arrector pili.
   B) cortex of the hair.
   C) cuticle of the hair.
   D) matrix cells.
   E) papilla of the hair.

   Answer: E

37) Production of new hairs is the responsibility of the
   A) arrector pili.
   B) cortex of the hair.
   C) cuticle of the hair.
   D) matrix cells.
   E) papilla of the hair.

   Answer: D

38) Sebaceous glands usually secrete their products into the
   A) blood.
   B) necks of hair follicles.
   C) peaks of epidermal ridges.
   D) melanosomes.
   E) external auditory canal.

   Answer: B

39) Fats, cholesterol, and pheromones are important components of
   A) keratin.
   B) melanin.
   C) sebum.
   D) eccrine sweat
   E) hair and nails.

   Answer: C

40) Which of the following is NOT a function of the skin?

A) absorption

B) protection

C) sensation

D) maintains homeostasis

Answer: A

41) Place the following in the order they would be severed by a knife during surgery:

1. stratum lucidum

2. stratum corneum

3. stratus basale

4. stratum granulosum

5. dermis

A) 1, 2, 3, 4, 5.

B) 3, 2, 4, 1, 5.

C) 4, 2, 1, 5, 3.

D) 2, 1, 4, 3, 5.

E) 5, 4, 3, 2, 1.

Answer: D

42) Vitamin D production by the skin depends on

A) exposure to ultraviolet radiation.

B) the presence of thyroid hormones.

C) the presence of keratin.

D) a normal body temperature.

E) All answers are correct.

Answer: A

43) ALL of the following events occur during deep wound healing EXCEPT

A) vasodilation of blood vessels.

B) suspension of the rules of contact inhibition.

C) formation of a blood clot.

D) synthesis of scar tissue by fibroblasts.

E) increased permeability of blood vessels.

Answer: B

44) Which of the following layers of epidermis is in a constant state of mitosis?

A) Stratum basale

B) Stratum spinosum

C) Stratum lucidum

D) Stratum granulosum

E) Both stratum basale and stratum spinosum.

Answer: A

45) The process of fibrosis results in

A) some form of skin cancer.

B) scar formation.

C) excessive production of skin pigment.

D) excessive production of hair and nails.

E) premature wrinkling of skin.

Answer: B

46) Blisters form in second-degree burns because

A) the epidermis and dermis separate, and tissue fluid accumulates between the layers.

B) the dermal papillae extend through the damaged epidermis and are exposed to the external environment.

C) damaged nerve endings swell.

D) accumulated tissue fluid is necessary for scar formation.

E) the cells of the stratum basale are reproducing at such a rapid rate.

Answer: A

47) The most common forms of skin cancer are all caused, at least in part, by

A) chronic dryness of skin.

B) over-secretion by sudoriferous glands.

C) chronic exposure to sunlight.

D) over-production of keratin.

E) chronically reduced blood flow in the dermis.

Answer: C

48) Sudoriferous glands are categorized as two distinct types. Which of the following are the two types?

A) eccrine and apocrine.

B) eccrine and sebaceous.

C) apocrine and sebaceous.

D) mammillary and ceruminous.

E) holocrine and mammillary

Answer: A

49) Differences in skin color among human races are due primarily to the
   A) total number of melanocytes.
   B) total number of keratinocytes.
   C) amount of melanin produced by melanocytes.
   D) amount of keratin produced by keratinocytes.
   E) amount of iron in hemoglobin molecules.

   Answer: C

50) During deep wound healing, mesenchyme cells that migrate to the site of injury during the inflammatory phase will develop into
   A) keratinocytes.
   B) melanocytes.
   C) fibroblasts.
   D) phagocytes.
   E) collagen fibers.

   Answer: C

MATCHING. Choose the item in column 2 that best matches each item in column 1.

51) Helps regulate body temperature.
52) UV light activates a precursor of this compound.
53) Darkly staining granules of the stratum granulosum.
54) A precursor to vitamin A.
55) Common form of baldness.

A. Vitamin D
B. melanin
C. alopecia
D. Eccrine sweat
E. Carotene

   Answers: 51) D.   52) A.   53) B.   54) E.   55) C.

56) Fibrous protein providing mechanical protection in the epidermis.
57) Provides protection from UV light.
58) Secreted during emotional stress and sexual excitement.
59) Pigment responsible for pinkish red color of "white skin".
60) Forms sticky barrier in external auditory canal.

A. Melanin
B. Cerumen
C. Apocrine sweat
D. Keratin
E. Hemoglobin

   Answers: 56) D.   57) A.   58) C.   59) E.   60) B.

61) Produce a protein that provides protection from mechanical injury and bacterial invasion.
62) Produce earwax.
63) Produce a substance that helps protect the body from UV light.
64) Function in sensation of touch.
65) Work with helper T cells to provide immunity.

A. Melanocytes
B. Langerhans cells
C. Ceruminous glands
D. Keratinocytes
E. Merkel cells

   Answers: 61) D.   62) C.   63) A.   64) E.   65) B.

66) Provide increased surface area for nutrient, waste, and gas exchange.

A. Sebaceous glands

67) Produce a product that helps prevent excessive evaporation of water from the skin.

B. Arrector pili

68) Produce a product that helps regulate body temperature.

C. Dermal papillae

69) Raises hair to vertical position.

D. Pacinian corpuscles

70) Function in sensations of pressure.

E. Sudoriferous glands

Answers: 66) C. 67) A. 68) E. 69) B. 70) D.

SHORT ANSWER. Write the word or phrase that best completes each statement or answers the question.

71) The red/brown/black pigment in skin that absorbs UV light is _____.

Answer: melanin

72) Individuals who do not get enough exposure to sunlight or who do not consume enough fortified milk may develop a deficiency of vitamin _____.

Answer: D

73) The pinkish red color of the skin of white people is due to the pigment _____ in red blood cells.

Answer: hemoglobin

74) The outermost layer of a hair is called the _____.

Answer: cuticle

75) Contraction of smooth muscle called _____ pulls a hair shaft perpendicular to the skin surface.

Answer: arrector pili

76) The more common type of sweat gland is the _____ gland.

Answer: eccrine

77) The single layer of continually reproducing cells in the epidermis is called the _____.

Answer: stratum basale

78) The most superficial layer of cells in the epidermis is called the _____.

Answer: stratum corneum

79) The protein in the outer layer of the epidermis that provides protection against mechanical injury and bacterial invasion is _____.

Answer: keratin

80) The main function of eccrine gland sweat is to _____.

Answer: regulate body temperature

81) The layer of the epidermis seen in thick skin that is NOT seen in thin skin is the _____.

Answer: stratum lucidum

82) The most common skin cancers are the _____.

Answer: basal cell carcinomas

83) The branch of medicine that specializes in diagnosing and treating skin disorders is _____.

Answer: dermatology

84) A yellowed appearance of skin and the whites of the eyes due to buildup of bilirubin resulting from liver disease is called _____.

Answer: jaundice

85) Redness of the skin due to increased blood flow is known as _____.

Answer: erythema

86) The most common of the cell types in the epidermis is the _____.

Answer: keratinocyte

87) Cells that arise from the red bone marrow and migrate to the epidermis where they participate in immune responses are the _____.

Answer: Langerhans cells

88) Cells in the epidermis that function in the sensation of touch are the _____.

Answer: Merkel cells

89) The stem cells of the epidermis are located in the stratum _____.

Answer: basale (germinativum)

90) Keratohyalin is a protein distinctive of the stratum _____.

Answer: granulosum

91) A lipid-rich, water-repellent secretion is produced by structures known as _____ within the keratinocytes of the stratum granulosum.

Answer: lamellar granules

92) The characteristic of epidermal cells that causes them to stop migrating during wound healing once they are touching other epidermal cells on all sides is called _____.

Answer: contact inhibition

93) The process of scar tissue formation is called _____.

Answer: fibrosis

94) When body temperature begins to fall, to prevent further heat loss blood vessels in the skin will _____.

Answer: constrict

95) In the negative feedback loop in which the integumentary system helps regulate body temperature, the effectors are the _____ and the _____.

Answer: sudoriferous glands; blood vessels in the skin

ESSAY. Write your answer in the space provided or on a separate sheet of paper.

96) List and briefly discuss the functions of skin.

Answer: 1. regulation of body temperature via sweat production and changes in blood flow

2. protection from mechanical injury, bacterial invasion (via keratin), dehydration (via product of lamellar granules), and UV light (via melanin)

3. sensory reception via receptors for temperature, touch, pressure, pain

4. excretion via sweat

5. immunity via Langerhans cells

6. blood reservoir for 8 minus 10% of total blood flow

7. synthesis of vitamin D from precursors in skin activated by UV light

97) Describe the structural characteristics of the epidermis that relate to its protection function.

Answer: Multiple layers of cells in stratified squamous epithelium help resist friction. Keratin of intermediate filaments provides strength to tissue by binding cells tightly together and to underlying tissue, and by producing a barrier to microbes. Lamellar granules of keratinocytes produce lipid-rich, water-repellent sealant to protect from dehydration and entry of foreign materials. Melanin, produced by melanocytes, protects underlying tissue from UV light. Sebum secreted onto the surface helps protect from dehydration and microbial invasion. Langerhans cells participate in immune reponse to microbial invasion.

98) John has just been brought into the emergency room following a fiery explosion at a chemical plant. He is diagnosed with third degree burns over the anterior surfaces of his arms and trunk. What specific structural damage has occurred to his skin? What risks to John's life have resulted from this damage?

Answer: John has lost approximately 36% of his skin's surface area (according to the Rule of Nines), which leads to severe systemic effects. The epidermis, dermis, and associated structures have been destroyed. Sensory function is lost. Loss of epidermis (and so, lost keratin and Langerhans cells) leaves John open to microbial invasion. Loss of keratinized structures and lamellar granules allows for extreme water loss, leading to dehydration, reduced blood volume and circulation, and decreased urine output.

99) Describe the process of deep wound healing.

Answer: 1. Inflammatory phase – blood clot forms to loosely unite wound edges; blood vessels dilate and become more permeable; phagocytes and mesenchyme cells migrate to site of injury.

2. Migratory phase – clot becomes scab; epithelial cells migrate beneath scab; fibroblasts synthesize scar tissue; begin repair of damaged blood vessels; granulation tissue fills wound.

3. Proliferative phase – extensive growth of epithelium beneath scab; fibroblasts deposit collagen fibers in random patterns; further vessel repair.

4. Maturation phase – scab falls off; collagen fibers become organized; fibroblasts decrease in number; blood vessels repaired.

100) Compare and contrast the locations and structure of thin and thick skin.

Answer: Thick skin is found on palms and palmar surfaces of digits and soles, while thin skin is found in all other areas but not these. Thick skin is 4-5X thicker than thin skin. The stratum lucidum is present in thick skin but not thin, and strata spinosum and corneum are thicker. Thick skin exhibits epidermal ridges, more sweat glands, and denser sensory receptors. Thin skin has hair follicles and sebaceous glands, while thick skin does not.

MULTIPLE CHOICE: Choose the one alternative that best completes the statement or answers the question.

1)   The process of hemopoiesis occurs in
   A) osteons.
   B) the periosteum.
   C) yellow bone marrow.
   D) red bone marrow.
   E) Both C and D are correct.

   Answer: D

2)   The primary function of yellow bone marrow is
   A) triglyceride storage.
   B) hemopoiesis.
   C) collagen production.
   D) to prevent collapse of trabeculae.
   E) to provide a blood supply to osteocytes in lacunae.

   Answer: A

3)   Hydroxyapatite is the
   A) combination of calcium compounds in bone matrix.
   B) collagen portion of bone matrix.
   C) concentric ring structure seen in compact bone.
   D) central canal of an osteon.
   E) hormone that stimulates breakdown of bone matrix.

   Answer: A

4)   During endochondral bone formation, the primary center of ossification forms in the
   A) proximal epiphysis.
   B) distal epiphysis.
   C) epiphyseal plate.
   D) diaphysis.
   E) metaphysis.

   Answer: D

5) In endochondral bone formation the original pattern for the bone is made of
   A) osseous tissue.
   B) keratin.
   C) dense irregular connective tissue.
   D) elastic cartilage.
   E) hyaline cartilage.

   Answer: E

6) The function of the epiphyseal plate is to
   A) allow more flexibility in a long bone.
   B) allow a means by which the bone can increase in diameter.
   C) allow a means by which the bone can increase in length.
   D) provide nourishment to isolated osteocytes.
   E) Both B and C are correct.

   Answer: C

7) The shaft of a long bone is the
   A) osteon.
   B) Haversian canal.
   C) metaphysis.
   D) diaphysis.
   E) epiphysis.

   Answer: D

8) The function of osteoblasts is to
   A) break down bone.
   B) produce blood cells.
   C) produce new bone formation.
   D) add new tissue to the periosteum.
   E) provide nourishment to the cells of the articular cartilage.

   Answer: C

9) Osteons are typical of the structure of
   A) compact bone.
   B) epiphyseal plates.
   C) spongy bone.
   D) the endosteum.
   E) All of these contain osteons.

   Answer: A

10) The epiphyseal plate is located in the

A) trabeculae.

B) metaphysis.

C) periosteum.

D) diaphysis.

E) epiphysis.

Answer: B

11) The distal and proximal extremities of a bone are the

A) lacunae.

B) trabeculae.

C) metaphysis.

D) diaphysis.

E) epiphysis.

Answer: E

12) Lamellae are the

A) plates of bone in spongy bone.

B) spaces in osteons in which osteocytes are located.

C) channels that contain cytoplasmic extensions of osteocytes in compact bone.

D) layers of bone in an osteon.

E) active cells in the epiphyseal plate.

Answer: D

13) Friction reduction and shock absorption are functions of what part of a long bone?

A) the articular cartilage

B) the periosteum

C) the epiphyseal plate

D) the bone marrow

E) the endosteum

Answer: A

14) Nutrients are provided to osteocytes in compact bone by

A) transport through canaliculi from blood vessels to the central canals.

B) blood in the marrow cavity.

C) blood seeping through the matrix of interstitial lamellae.

D) dissolving the matrix around them via enzymes in the lacunae.

E) blood in yellow bone marrow.

Answer: A

15) The endosteum is the
    A) covering of a long bone.
    B) marrow in spongy bone.
    C) layer of active chondrocytes in the epiphyseal plate.
    D) end of a long bone.
    E) lining of the medullary cavity.

    Answer: E

16) When an osteoblast becomes isolated in a lacuna, it is transformed into a(n)
    A) macrophage.
    B) osteocyte.
    C) osteoclast.
    D) fibroblast.
    E) osteal fiber.

    Answer: B

17) The tiny channels connecting osteocytes with the central canal of an osteon are called
    A) Haversian canals.
    B) lamellae.
    C) canaliculi.
    D) lacunae.
    E) perforating (Volkmann's) canals.

    Answer: C

18) The hormone produced by the thyroid gland that lowers serum calcium levels is
    A) IGF.
    B) PTH.
    C) calcitonin.
    D) growth hormone.

    Answer: C

19) Which of the following contains bone marrow?
    A) bone matrix
    B) epiphyseal cartilage
    C) periosteum
    D) medullary cavity

    Answer: D

20) The hormone from the parathyroid gland functions to
   A) stimulate the activity of the osteoblasts.
   B) stimulate the activity of the osteoclasts.
   C) cause the level of blood calcium to decrease.
   D) cause the level of lead in the bone to increase.

   Answer: A

21) During endochondral ossification, calcification of cartilage matrix causes the death of chondrocytes. This leads to
   A) death of the developing bone.
   B) hemopoiesis.
   C) closing of the epiphyseal plates.
   D) erosion of articular cartilage.
   E) development of the marrow cavity.

   Answer: E

22) Osteogenic cells are located in
   A) the inner portion of the periosteum.
   B) lacunae at the edge of lamellae in osteons.
   C) the epiphyseal plate.
   D) the articular cartilage.
   E) All of these are correct.

   Answer: A

23) Collagen is secreted by
   A) mesenchyme cells.
   B) osteoblasts.
   C) osteoclasts.
   D) hydroxyapatites.
   E) Both B and C are correct.

   Answer: B

24) If excessive growth hormone is secreted after the epiphyseal line has closed, a condition known as____ may result.
   A) acromegaly
   B) pituitary giantism
   C) achondroplasia
   D) pituitary dwarfism

   Answer: A

25) ALL of the following are normal sites of hemopoiesis EXCEPT
   A) clavicle (collarbone).
   B) hip bone.
   C) rib.
   D) sternum.
   E) vertebral bodies.

   Answer: A

26) Osteoclasts are located primarily in the
   A) lining of the central canal.
   B) lacunae at the edges of lamellae.
   C) epiphysis.
   D) endosteum.
   E) epiphyseal plate.

   Answer: D

27) Hydroxyapatite consists mostly of
   A) calcium phosphate.
   B) collagen.
   C) magnesium hydroxide.
   D) dense irregular connective tissue.
   E) hyaline cartilage.

   Answer: A

28) Growth in bone length is primarily a function of the
   A) metaphysis.
   B) diaphysis.
   C) endosteum.
   D) periosteum.
   E) epiphysis.

   Answer: E

29) The tensile strength of bone is provided by
   A) the periosteum.
   B) mineral salts.
   C) collagen fibers.
   D) hyaline cartilage.
   E) extensions of osteocytes into canaliculi.

   Answer: C

30) Which of the following best describes a compound (open) fracture?
   A) The bone is broken in more than one place.
   B) The broken ends of the bone protrude through the skin.
   C) The bone is usually twisted apart.
   D) The fracture is at right angles to the long axis of the bone.
   E) One bone fragment is driven into the other.

Answer: B

31) Blood vessels, lymphatic vessels, and nerves from the periosteum penetrate compact bone via
   A) central (Haversian) canals.
   B) perforating (Volkmann's) canals.
   C) canaliculi.
   D) nutrient foramina.
   E) circumferential lamellae.

Answer: B

32) The function of the nutrient foramina of a long bone is to provide a passageway for
   A) blood vessels through the center of osteons.
   B) ionized calcium ions to leave the bone matrix.
   C) hormones to enter the bone matrix.
   D) osteoblasts to reach the sites of new bone growth.
   E) blood vessels into the medullary cavity.

Answer: E

33) A fracture in the shaft of a bone is a break in the
   A) epiphysis.
   B) periosteum.
   C) diaphysis.
   D) articular cartilage.

Answer: C

34) ALL of the following are TRUE regarding epiphyseal plates EXCEPT
   A) They "close" between the ages of 18 and 20.
   B) They are avascular.
   C) They get thinner and thinner as a child ages.
   D) Cartilage is replaced by bone on the diaphyseal side of the plate.
   E) They consist of hyaline cartilage in four zones.

Answer: C

35) Bone grows in diameter by
   A) appositional growth on the periosteal side of the bone.
   B) interstitial growth.
   C) activity at the epiphyseal plate.
   D) activity of the hemopoietic tissue.
   E) A, B, and C are all correct.

Answer: A

36) ALL of the following bones would be produced via endochondral ossification EXCEPT the
   A) tibia.
   B) femur.
   C) frontal.
   D) humerus.
   E) radius.

Answer: C

37) The role of Vitamin C in bone growth is that it
   A) promotes absorption of calcium from the intestinal tract.
   B) is required for collagen synthesis.
   C) stimulates osteoclast activity.
   D) promotes secretion of calcitonin.
   E) acts as an insulin-like growth factor.

Answer: B

38) The role of human growth hormone in bone growth is that it
   A) stimulates production of osteoclasts in the endosteum.
   B) triggers calcification of matrix at the epiphyseal plate.
   C) promotes formation of calcitriol.
   D) stimulates production of IGFs.
   E) inhibits osteoclast activity.

Answer: D

39) Growth at the epiphyseal plate is shut down by
   A) Vitamin A.
   B) human growth hormone.
   C) calcitonin.
   D) PTH.
   E) estrogens.

Answer: E

40) Giantism results from childhood oversecretion of

A) hGH.

B) PTH.

C) calcitonin.

D) Oversecretion of any of these results in giantism.

Answer: A

41) In both intramembranous and endochondral ossification, the first stage of development of bone is

A) penetration of the diaphysis by a nutrient artery.

B) migration of mesenchyme cells to the area of bone formation.

C) formation of a cartilage model of the bone.

D) fusion of trabeculae.

E) development of a periosteal bud.

Answer: B

42) The cells responsible for maintaining the daily cellular activities of bone tissue, such as exchange of nutrients and wastes with the blood, are the

A) osteogenic cells.

B) osteoblasts.

C) osteoclasts.

D) osteocytes.

E) chondroblasts.

Answer: D

43) The crystallization of mineral salts in bone occurs only in the presence of

A) collagen.

B) PTH.

C) osteoclasts.

D) hyaline cartilage.

E) yellow bone marrow.

Answer: A

44) Promotion of apoptosis of osteoclasts is an effect of

A) PTH.

B) hGH.

C) Vitamin C.

D) hydroxyapatite.

E) estrogens.

Answer: E

45) Bone constantly remodels and redistributes its matrix along lines of
   A) blood flow.
   B) nervous stimulation.
   C) canaliculi.
   D) mechanical stress.
   E) overlying muscles.

   Answer: D

46) A bedridden person loses bone mass because of
   A) decreased calcitonin activity due to immobility.
   B) lack of sufficient mechanical stress on bones.
   C) increased conversion of monocytes to osteoclasts.
   D) pressure on the kidney that leads to reduced calcitriol production.
   E) insufficient blood flow to bones.

   Answer: B

47) ALL of the following are effects of parathyroid hormone EXCEPT
   A) increased protein synthesis.
   B) increased numbers and activity of osteoclasts.
   C) reabsorption of calcium ions by the kidney.
   D) elimination of phosphate ions by the kidney.
   E) promotion of calcitriol production by the kidney.

   Answer: A

48) The formation of the crystalline structure of bone is termed
   A) crystallization.
   B) deposition.
   C) calcification.
   D) ossification.
   E) fibrosis.

   Answer: D

49) The "growth spurt" of teenage years is due to an increase in the level of
   A) hGH.
   B) sex steroids.
   C) PTH.
   D) calcitonin.

   Answer: B

50) Bones are more brittle in the elderly because
    A) levels of calcitonin are higher in the elderly.
    B) there is less collagen relative to the amount of mineral salts.
    C) low levels of growth hormone prevent deposition of calcium in bone.
    D) the calcium salts in the bone matrix are of a different type than those in younger people.
    E) too much collagen takes up space that should be occupied by hydroxyapatites.

    Answer: B

MATCHING. Choose the item in column 2 that best matches each item in column 1.

51) Develop into osteoblasts.
52) Form the matrix of cartilage.
53) Sit in lacunae in bone matrix.
54) Do not undergo mitosis; secrete collagen.
55) Located on bone surfaces; function in bone resorption.

A. Osteoblasts
B. Osteocytes
C. Chondroblasts
D. Osteogenic cells
E. Osteoclasts

Answers: 51) D.   52) C.   53) B.   54) A.   55) E.

56) Thin plates of bone in spongy bone.
57) Stem cells from which more differentiated bone and cartilage cells arise.
58) The mineral portion of bone matrix.
59) Units of concentric lamellae in compact bone.
60) Tiny channels that allow communication between osteocytes and blood vessels in central canals.

A. Trabeculae
B. Osteons
C. Canaliculi
D. Mesenchyme cells
E. Hydroxyapatites

Answers: 56) A.   57) D.   58) E.   59) B.   60) C.

61) Degeneration of articular cartilage due to overuse or aging.
62) Bone cancer primarily affecting osteoblasts.
63) Condition in whichs bone resorption outpaces bone deposition.
64) Condition in children in which bones fail to calcify, resulting in bone deformities.
65) An infection of bone often caused by Staphylococcus.

A. Osteogenic sarcoma
B. Rickets
C. Osteoporosis
D. Osteoarthritis
E. Osteomyelitis

Answers: 61) D.   62) A.   63) C.   64) B.   65) E.

66) Fracture of the distal end of the fibula fracture.

67) Broken ends of bone protrude through skin.

68) Bone has splintered at the site of impact.

69) Fracture of the distal end of the radius.

70) Microscopic fissures resulting from repeated, strenuous activities.

A. Stress fracture
B. Pott's
C. Compound fracture
D. Colles' fracture
E. Comminuted fracture

Answers: 66) B.   67) C.   68) E.   69) D.   70) A.

SHORT ANSWER. Write the word or phrase that best completes each statement or answers the question.

71) Destruction of matrix by osteoclasts is called bone _____.

Answer: resorption

72) A partial fracture in which one side of the bone is broken and the other side bends is known as a(n) _____ fracture.

Answer: greenstick

73) The membrane that lines the medullary cavity of a long bone is called the _____.

Answer: endosteum

74) Mature bone cells that are completely surrounded by matrix are called _____.

Answer: osteocytes

75) A fracture of the distal end of the radius in which the distal fragment is displaced posteriorly is called a(n) _____ fracture.

Answer: Colles'

76) The channels in osteons that connect lacunae with central canals are called _____.

Answer: canaliculi

77) The granulation tissue of a healing bone fracture is called a(n) _____.

Answer: procallus

78) In intramembranous ossification, the highly vascularized mesenchyme on the outside of the new bone develops into the _____.

Answer: periosteum

79) A series of microscopic fissures in bone that result from repeated, strenuous activities is known as a(n) _____ fracture.

Answer: stress

80) The part of a long bone that is not covered by periosteum is covered by _____.

Answer: articular cartilage (hyaline cartilage)

81) Cells whose primary function is bone resorption are the _____.

Answer: osteoclasts

82) The process by which the fractured ends of a bone are brought into alignment is called _____.

Answer: reduction

83) Bone is the major reservoir for the minerals _____ and _____.

Answer: calcium; phosphorus

84) The two most important hormones involved in the regulation of calcium ion levels in the blood are _____ and _____.

Answer: parathyroid hormone; calcitonin

85) Increasing the activity of osteoclasts is one of the major effects of the _____.

Answer: parathyroid hormone

86) Blood vessels run longitudinally through compact bone in spaces known as _____.

Answer: central (Haversian) canals

87) Thin plates of bone in spongy bone are called _____.

Answer: trabeculae

88) The flat bones of the skull form by _____ ossification.

Answer: intramembranous

89) The flexible rod of mesoderm that defines the midline of the embryo and gives it some rigidity is the _____.

Answer: notochord

90) The process by which blood cells are produced in red bone marrow is called _____.

Answer: hemopoiesis

91) The main function of yellow bone marrow is _____.

Answer: triglyceride storage

92) The shaft of a long bone is called the _____.

Answer: diaphysis

93) Bone constantly remodels and redistributes matrix along lines of _____.

Answer: mechanical stress

94) The dense irregular connective tissue that surrounds bone surfaces not covered by articular cartilage is the _____.

Answer: periosteum

95) Levels of calcium ions in the blood are decreased by the effects of the hormone _____.

Answer: calcitonin

ESSAY. Write your answer in the space provided or on a separate sheet of paper.

96) Patient X has a tumor of the parathyroid glands that causes a hypersecretion from these glands. Predict the effect on the skeletal system and on the secretion of calcitonin.

Answer: High levels of PTH would cause high levels of osteoclast activity, thus removing calcium from bones. Bones would become weak and soft. Excess phosphate would be lost from the kidneys. High levels of calcium ions in blood may disrupt nerve and muscle function. Calcitonin levels would probably be high, trying to restore homeostasis by increasing deposition of calcium into bone.

97) Describe the signs and symptoms of osteoporosis and describe the risk factors for developing osteoporosis.

Answer: In osteoporosis, bone resorption outpaces bone deposition so that bone mass is depleted, sometimes to the point of spontaneous fracture. Pain and height loss may occur as vertebrae shrink. Postmenopausal women are especially at risk due to lower initial bone mass than men due to dramatically reduced estrogen levels. Family history may play a role, as does ethnicity (white and Asian women have a higher rate of disease), inactivity, cigarette smoking, excessive alcohol consumption, and a diet low in calcium and vitamin D.

98) Archaeologists have unearthed the upper arm bones of some ancient humans. They note with interest that these bones have greatly enlarged deltoid tuberosities, which are the points where the deltoid (shoulder) muscles attach to the lateral surfaces of these bones. What might they infer from this finding about the life and activities of these people, and why?

Answer: Possibly some activity that required the repeated use of this muscle (such as archery or rowing), and that caused continual mechanical stress on this point of muscle attachment, thus inducing remodeling via addition of bone.

99) Describe the process by which bone increases in length and diameter.

Answer: The only means by which bone can increase in length is by activity at the epiphyseal plate. Until full height is reached, the plate consists of layers of chondrocytes that generate matrix, which is then calcified and replaced by bone matrix secreted by osteoblasts on the diaphyseal side of the plate. Around ages 18-20 the cartilage is replaced completely by bone and no more lengthwise growth can occur. Bone increases in diameter via appositional growth as new bone matrix is laid down by osteoblasts in the periosteum.

100) Describe the process by which bone increases in length and diameter.

Answer: The only means by which bone can increase in length is by activity at the epiphyseal plate. Until full height is reached, the plate consists of layers of chondrocytes which generate matrix which is then calcified and replaced by bone matrix secreted by osteoblasts on the diaphyseal side of the plate.

CHAPTER 7   The Skeletal System: The Axial System

MULTIPLE CHOICE. Choose the one alternative that best completes the statement or answers the question.

1) The auditory meatus is located in the
   A) frontal bone.
   B) temporal bone.
   C) parietal bone.
   D) occipital bone.
   Answer: B

2) Which of the following best defines a short bone?
   A) a bone with greater length than width
   B) a bone consisting of two parallel plates of compact bone
   C) a bone whose length and width are nearly equal
   D) a bone in a suture
   E) a bone measuring not more than five centimeters in length
   Answer: C

3) Which of the following best defines a flat bone?
   A) any bone in the skull
   B) a bone consisting of two parallel plates of compact bone
   C) a bone whose length and width are nearly equal
   D) a bone that has no curves in its structure
   E) a bone in a suture
   Answer: B

4) The carpals and tarsal are classified as
   A) flat bones.
   B) irregular bones.
   C) Wormian bones.
   D) short bones.
   E) long bones.
   Answer: D

5) The scapula is an example of a(n)
   A) flat bone.
   B) irregular bone.
   C) Wormian bone.
   D) short bone.
   E) long bone.

   Answer: A

6) The patellae are examples of
   A) flat bones.
   B) irregular bones.
   C) Wormian bones.
   D) short bones.
   E) sesamoid bones.

   Answer: E

7) ALL of the following are part of the axial skeleton EXCEPT the
   A) occipital bone.
   B) hyoid bone.
   C) vertebrae.
   D) coxal bones.
   E) sternum.

   Answer: D

8) Which of the following bones is considered to be sesamoid bone?
   A) humerus
   B) coxal
   C) hyoid
   D) patella
   E) talus

   Answer: C

9) The only movable bone of the skull is the
   A) temporal.
   B) maxilla.
   C) mandible.
   D) zygomatic.
   E) nasal.

   Answer: C

10) Which of the following sutures generally does NOT persist into adulthood?
   A) metopic
   B) sagittal
   C) coronal
   D) lambdoid
   E) squamous

   Answer: A

11) The bones forming the greater portions of the sides and roof of the cranial cavity are the
   A) frontals.
   B) temporals.
   C) sphenoids.
   D) occipitals.
   E) parietals.

   Answer: E

12) ALL of the following terms refer to processes or projections from bones EXCEPT
   A) tubercle.
   B) fossa.
   C) trochanter.
   D) condyle.
   E) spine.

   Answer: B

13) The zygomatic process is part of the
   A) sphenoid bone.
   B) frontal bone.
   C) temporal bone.
   D) zygomatic bone.
   E) maxilla.

   Answer: C

14) The internal auditory meatus is the opening through which
   A) cranial nerves VII and VIII pass.
   B) sound waves are directed into the ear.
   C) the ossicles vibrate.
   D) air passes to enter the middle ear.
   E) ligaments suspending the hyoid bone pass.

   Answer: A

15) ALL of the following are considered part of the appendicular skeleton EXCEPT the
   A) humerus.
   B) coxal bones.
   C) fibula.
   D) ribs.
   E) calcaneus.

   Answer: D

16) There are normally TWO of EACH of the following bones EXCEPT the
   A) vomer.
   B) maxilla.
   C) nasal.
   D) temporal.
   E) zygomatic.

   Answer: A

17) The carotid artery passes through the carotid foramen in the
   A) greater wings of the sphenoid.
   B) cribriform plate of the ethmoid.
   C) mastoid process of the temporal bone.
   D) petrous portion of the temporal bone.
   E) occipital condyles.

   Answer: D

18) The ligamentum nuchae is a fibroelastic ligament extending between the seventh cervical vertebra and the
   A) external occipital protuberance.
   B) mastoid processes of the temporal bone.
   C) coccyx.
   D) dens of the axis.
   E) crista galli of the ethmoid.

   Answer: A

19) The superior orbital fissure is located
   A) in the supraorbital margin.
   B) between the anterior aspects of the greater and lesser wings of the sphenoid bone.
   C) between the petrous portion of the temporal bone and the occipital bone.
   D) in the orbit between the sphenoid and ethmoid bones.
   E) between the lacrimal and nasal bones.

   Answer: B

20) Some muscles that move the mandible attach to the
    A) sella turcica of the sphenoid bone.
    B) styloid processes of the temporal bone.
    C) external occipital protuberance.
    D) perpendicular plate of the ethmoid.
    E) pterygoid processes of the sphenoid bone.

    Answer: E

21) The internal and middle ear are housed by the
    A) external auditory meatus.
    B) mastoid process of the temporal bone.
    C) sella turcica of the sphenoid bone.
    D) petrous portion of the temporal bone.
    E) greater wings of the sphenoid bone.

    Answer: D

22) The mandibular fossa of the temporal bone articulates with what part of the mandible?
    A) condylar process
    B) coronoid process
    C) alveolar process
    D) zygomatic process
    E) mental foramen

    Answer: A

23) The temporal bone articulates with ALL of the following EXCEPT the
    A) parietal bone.
    B) zygomatic bone.
    C) mandible.
    D) frontal bone.
    E) sphenoid bone.

    Answer: D

24) Directly anterior to the sphenoid bone and posterior to the nasal bones is the
    A) zygomatic bone.
    B) palatine bone.
    C) maxilla.
    D) hyoid.
    E) ethmoid bone.

    Answer: E

25) The hyoid bone is suspended from the
    A) mastoid processes of the temporal bones.
    B) occipital condyles.
    C) superior nasal conchae of the ethmoid bones.
    D) sella turcica of the sphenoid bone.
    E) styloid processes of the temporal bones.

    Answer: E

26) The superior articular facets of the atlas articulate with the
    A) occipital condyles.
    B) mastoid processes of the temporal bones.
    C) dens of the axis.
    D) inferior articular facets of the axis.
    E) first ribs.

    Answer: A

27) Which of the following is located between the greater wing of the sphenoid bone and the maxilla?
    A) optic foramen
    B) infraorbital foramen
    C) supraorbital foramen
    D) superior orbital fissure
    E) inferior orbital fissure

    Answer: E

28) Cleft palate results from incomplete fusion of the
    A) vomer and perpendicular plate.
    B) alveolar processes of the maxillae.
    C) palatine processes of the maxillae.
    D) pterygoid processes of the sphenoid.
    E) nasal bones.

    Answer: C

29) Tears pass into the nasal cavity via the
    A) mental foramen.
    B) jugular foramen.
    C) carotid foramen.
    D) infraorbital foramen.
    E) lacrimal fossa.

    Answer: E

30) The pituitary gland is located in the
   A) sella turcica of the sphenoid bone.
   B) petrous portion of the temporal bone.
   C) mastoid process of the temporal bone.
   D) hypoglossal canal of the occipital bone.
   E) crista galli of the ethmoid bone.

   Answer: A

31) Which of the following are NOT part of the ethmoid bone?
   A) superior nasal conchae
   B) middle nasal conchae
   C) inferior nasal conchae
   D) lateral masses
   E) olfactory foramina

   Answer: C

32) The bone whose superior border articulates with the perpendicular plate of the ethmoid to form the nasal septum is the
   A) lacrimal.
   B) nasal.
   C) palatine.
   D) sphenoid.
   E) vomer.

   Answer: E

33) Which of the following has both an alveolar process and a coronoid process?
   A) mandible
   B) maxilla
   C) zygomatic
   D) vomer
   E) palatine

   Answer: A

34) The function of the nasal conchae is to
   A) provide a surface for muscle attachment.
   B) create turbulence in inspired air for cleansing purposes.
   C) provide extensive surface area for gas exchange.
   D) act as an anchor for periodontal ligaments.
   E) protect olfactory nerves as they travel to the brain.

   Answer: B

35) Paranasal sinuses are found in ALL of the following bones EXCEPT the
    A) frontal.
    B) sphenoid.
    C) zygomatic.
    D) ethmoid.
    E) maxilla.

    Answer: C

36) ALL of the following form part of the orbit EXCEPT the
    A) zygomatic.
    B) maxilla.
    C) sphenoid.
    D) vomer.
    E) lacrimal.

    Answer: D

37) The nasal septum is formed by cartilage and the
    A) pterygoid processes and perpendicular plate of the sphenoid.
    B) horizontal plate of the palatine and the perpendicular plate of the ethmoid.
    C) vomer and perpendicular plate of the ethmoid.
    D) vomer and perpendicular plate of the sphenoid.
    E) nasal bones.

    Answer: C

38) The tongue is supported by the
    A) hyoid.
    B) maxilla.
    C) zygomatic.
    D) vomer.
    E) palatines.

    Answer: A

39) The spinal cord passes through the
    A) intervertebral foramen.
    B) transverse foramen.
    C) vertebral foramen.
    D) centrum.
    E) Both C and D are correct.

    Answer: C

40) When whiplash injuries result in death, the usual cause is damage to the medulla oblongata of the brain by the
   A) rupturing of intervertebral discs.
   B) dens of the axis.
   C) spinous processes of the cervical vertebrae.
   D) vertebra prominens.
   E) lateral masses of the atlas.

   Answer: B

41) The meninges that cover the brain attach anteriorly to the
   A) sella turcica of the sphenoid bone.
   B) styloid processes of the temporal bones.
   C) external occipital protuberance.
   D) perpendicular plate of the ethmoid.
   E) crista galli of the ethmoid bone.

   Answer: E

42) The laminae of a vertebra are the parts
   A) through which spinal nerves pass.
   B) that form the posterior portion of the vertebral arch.
   C) that form the transverse processes.
   D) to which the ribs attach.
   E) to which the intervertebral discs attach.

   Answer: B

43) Transverse foramina are seen in
   A) cervical vertebrae.
   B) thoracic vertebrae.
   C) lumbar vertebrae.
   D) sacral vertebrae.
   E) Both A and B are correct.

   Answer: A

44) ALL of the following foramina are in the sphenoid bone EXCEPT the
   A) foramen rotundum.
   B) superior orbital fissure.
   C) foramen ovale.
   D) optic foramen.
   E) jugular foramen.

   Answer: E

45) Anesthetic agents used in caudal anesthesia are sometimes injected into the
   A) lumbosacral joint.
   B) sacral hiatus.
   C) coccygeal cornua.
   D) nucleus pulposus.
   E) sacral ala.

   Answer: B

46) The xiphoid process is part of the
   A) atlas.
   B) axis.
   C) sternum.
   D) sacrum.
   E) coccyx.

   Answer: C

47) Herniated discs occur most often in which vertebral region?
   A) cervical
   B) thoracic
   C) lumbar
   D) sacral
   E) coccygeal

   Answer: C

48) Which of the following lists the regions of the vertebral column in the correct order from superior to inferior?
   A) cervical, lumbar, thoracic, coccygeal, sacral
   B) coccygeal, sacral, lumbar, thoracic, cervical
   C) coccygeal, lumbar, sacral, thoracic, cervical
   D) cervical, thoracic, lumbar, sacral, coccygeal
   E) thoracic, cervical, lumbar, sacral, coccygeal

   Answer: D

49) An exaggeration of the thoracic curve of the vertebral column is called
   A) lordosis.
   B) scoliosis.
   C) kyphosis.
   D) thoracic spine stenosis.

   Answer: C

50) The glossopharyngeal, vagus, and accessory nerves pass through the

   A) foramen rotundum.

   B) foramen lacerum

   C) carotid foramen.

   D) hypoglossal canal.

   E) jugular foramen.

   Answer: E

MATCHING: Choose the item in column 2 that best matches each item in column 1.

51) Sella turcica.   A. Mandible

52) Styloid process.   B. Sphenoid

53) Condylar process.   C. Zygomatic

54) Superior nuchal line.   D. Temporal

55) Temporal process.   E. Occipital

   Answers: 51) B. 52) D. 53) A. 54) E. 55) C.

56) Manubrium.   A. Palatine

57) Supraorbital margin.   B. Sternum

58) Palatine process.   C. Frontal

59) Crista galli.   D. Maxilla

60) Horizontal plate.   E. Ethmoid

   Answers: 56) B. 57) C. 58) D. 59) E. 60) A.

61) A smooth, flat surface.   A. Foramen

62) Part of a bone that forms an angle with the main body of a bone.   B. Fossa

63) An opening through which blood vessels, nerves or ligaments pass.   C. Facet

64) A small, rounded process.   D. Tubercle

65) A depression in or on a bone.   E. Ramus

   Answers: 61) C. 62) E. 63) A. 64) D. 65) B.

66) A tubelike passageway running within a bone.   A. Sulcus

67) A groove that accommodates a soft structure.   B. Fissure

68) A large, rounded, roughened process.   C. Crest

69) A prominent border or ridge.   D. Meatus

70) A narrow, cleftlike opening between adjacent parts of a bone.   E. Tuberosity

   Answers: 66) D. 67) A. 68) E. 69) C. 70) B.

SHORT ANSWER. Write the word or phrase that best completes each statement or answers the question.

71) There are _____ pairs of vertebrosternal ribs.

Answer: seven

72) The spaces between the ribs are called _____ spaces.

Answer: intercostal

73) An exaggeration of the thoracic curve of the vertebral column commonly seen in women with advanced osteoporosis is called _____.

Answer: kyphosis

74) The first cervical vertebra is called the _____.

Answer: atlas

75) Spinal nerves pass through openings between the vertebrae called _____.

Answer: intervertebral foramina

76) Bones located in tendons are referred to as _____ bones.

Answer: sesamoid

77) The passageway through the temporal bone that directs sound waves into the ear is the _____.

Answer: external auditory meatus

78) The suture located between the parietal bones and the occipital bone is the _____ suture.

Answer: lambdoid

79) The hyoid bone is suspended by ligaments from the _____ of the temporal bones.

Answer: styloid processes

80) Ribs that attach anteriorly to the cartilage of other ribs are referred to as false ribs or _____ ribs.

Answer: vertebrochondral

81) The internal and middle ear are housed within the _____ of the temporal bone.

Answer: petrous portion

82) When an infant begins to hold its head up, its vertebral column begins to develop a _____ curve.

Answer: cervical

83) The occipital condyles articulate with the _____.

Answer: atlas

84) The foramen magnum is a large hole in the _____ bone.

Answer: occipital

85) The major supporting structure of the nasal cavity is the _____ bone.

Answer: ethmoid

86) The olfactory foramina are located in the _____ of the ethmoid bone.

Answer: cribriform plate

87) The nasal septum is formed by the _____ and the _____.

Answer: perpendicular plate of the ethmoid; vomer

88) The triangular point of the ethmoid that serves as a point of attachment for the meninges is the _____.

Answer: crista galli

89) Two foramina in the mandible that serve as important sites for injection of dental anesthetics are the mandibular foramen and the _____ foramen.

Answer: mental

90) The superior and middle conchae are part of the _____ bone.

Answer: ethmoid

91) The adult vertebral column consists of _____ cervical vertebrae, _____ thoracic vertebrae, _____ lumbar vertebrae, _____ sacral vertebrae fused into one bone, and _____ coccygeal vertebrae fused into one or two bones.

Answer: 7; 12; 5; 5; 4

92) The hard palate is formed by the _____ of the _____ , and the _____ of the _____.

Answer: palatine processes; maxillae; horizontal plates; palatine bones

93) The prominences of the cheeks are formed by the _____ bones.

Answer: zygomatic

94) The tiny bones just posterior and lateral to the nasal bones are the _____.

Answer: lacrimal bones

95) The condylar process of the mandible articulates with the mandibular fossa of the _____.

Answer: temporal bone

ESSAY. Write your answer in the space provided or on a separate sheet of paper.

96) Describe the location and anatomical features of the sphenoid bone.

Answer: The bat-shaped sphenoid bone forms much of the cranial floor and articulates anteriorly with the frontal, laterally with the temporal, and posteriorly with the occipital. The body, containing the sphenoidal sinus, is centrally located between the ethmoid and the occipital. The sella turcica, which protects the pituitary, sits atop the body. The greater wings form the anterolateral cranial floor and part of the lateral wall of the skull. The lesser wings, anterior and superior to the greater wings, form the posterior part of the orbit. Foramina include the optic foramen, foraman ovale, foramen rotundum, and superior and inferior orbital fissures. The pterygoid processes project inferiorly from the junction of the body and the greater wings.

97) Where are the turbinates located and what is their function?

Answer: Superior and middle nasal conchae (turbinates) project like scrolls from the lateral masses of the ethmoid. The inferior turbinates are separate bones. All form the lateral walls of the nasal cavity. The scrolled structure forms a mucus-coated surface area that traps inhaled particles and warms and moistens air.

98) Where is the hyoid bone located? What is its function? What is the significance of finding a broken hyoid on autopsy?

Answer: The hyoid bone is suspended by ligaments and muscles from the styloid processes of the temporal bones. It is located in the anterior neck between the mandible and the larynx. It supports the tongue, providing attachment sites for some tongue muscles and for muscles of the neck and pharynx. Crushing injury to the neck, such as would occur in strangulation, is indicated by a broken hyoid.

99) Describe the classification of ribs based on their articulations with other bones.

Answer: True ribs (vertebrosternal) are pairs 1–7 and attach directly to the sternum via costal cartilages. False ribs (vertebrochondral) are pairs 8–10 and attach directly to each other's cartilage then to the cartilage of pair 7. False ribs (vertebral) also include pairs 11–12 which have no anterior attachment.

100) Name and describe the abnormal curvatures of the spine. How does each commonly develop?

Answer:
1. Scoliosis – abnormal lateral bending on the thoracic region; congenital, malformed vertebrae, chronic sciatica, one-sided paralysis, poor posture, one leg shorter than the other.
2. Kyphosis – exaggerated thoracic curvature; TB of spine or osteoporosis leading to collapse of vertebral bodies; also rickets, poor posture, geriatric disc degeneration.
3. Lordosis – exaggerated lumbar curvature; increased abdominal weight, poor posture, rickets, TB of spine.

CHAPTER 8   The Skeletal System: The Appendicular Skeleton

MULTIPLE CHOICE. Choose the one alternative that best completes the statement or answers the question.

1) ALL of the following are part of the appendicular skeleton EXCEPT the
   A) scapula.
   B) coxal bones.
   C) sternum.
   D) fibula.
   E) radius.

   Answer: C

2) Most of the structural differences between the male and female skeletons are related to adaptation for
   A) problems associated with height differences.
   B) problems associated with weight differences.
   C) hunting versus gathering.
   D) pregnancy and childbirth.
   E) roles in sexual activity.

   Answer: C

3) ALL of the following are typical of a female pelvis EXCEPT
   A) oval obturator foramen.
   B) acetabulum faces anteriorly.
   C) less movable coccyx.
   D) wider greater sciatic notch.
   E) shorter, wider sacrum.

   Answer: C

4) The coronoid and olecranon fossae are depressions found on the
   A) ulna.
   B) radius.
   C) scapula.
   D) humerus.
   E) femur.

   Answer: D

5) Most of the muscles of the forearm attach to the
   A) greater and lesser tubercles of the humerus.
   B) coronoid and olecranon fossae of the humerus.
   C) deltoid tuberosity of humerus.
   D) acromion and coracoid processes of the scapula.
   E) medial and lateral epicondyles of the humerus.

   Answer: E

6) The proximal end of the femur articulates with the
   A) acetabulum.
   B) patella.
   C) obturator foramen.
   D) condyles of the tibia.
   E) Both B and D are correct.

   Answer: A

7) Which of the following is part of the elbow joint?
   A) trochlear notch of the ulna
   B) ulnar notch of the radius
   C) glenoid fossa of the humerus
   D) styloid process of the radius
   E) head of the ulna

   Answer: A

8) ALL of the following are in the proximal row of carpal bones EXCEPT the
   A) hamate.
   B) lunate.
   C) scaphoid.
   D) pisiform.
   E) triquetrum.

   Answer: A

9) ALL of the following are in the distal row of carpal bones EXCEPT the
   A) trapezium.
   B) trapezoid.
   C) triquetrum.
   D) hamate.
   E) capitate.

   Answer: C

10) Which of the following is on the anterior side of the femur?
   A) gluteal tuberosity
   B) linea aspera
   C) intertrochanteric line
   D) intertrochanteric crest
   E) Both B and C are correct

   Answer: D

11) Genu varum is
   A) a landmark of the pelvic brim.
   B) the medical term for bowleggedness.
   C) a part of the carpal tunnel.
   D) the junction between the tibia and talus.
   E) the imaginary line the baby's head follows through the pelvis.

   Answer: B

12) The tibial tuberosity is the attachment point for the
   A) interosseous membrane.
   B) biceps brachii.
   C) patellar ligament.
   D) deltoid.
   E) flexor retinaculum.

   Answer: C

13) The bone that is considered a sesamoid bone is the
   A) clavicle.
   B) scaphoid.
   C) cuboid.
   D) patella.
   E) All are sesamoid bones except A.

   Answer: D

14) The base of the patella is the
   A) superior end.
   B) inferior end.
   C) anterior surface.
   D) posterior surface.
   E) lateral edge.

   Answer: A

15) The articular facets of the patella articulate with the
    A) medial and lateral epicondyles of the femur.
    B) medial and lateral condyles of the femur.
    C) medial and lateral epicondyles of the tibia.
    D) tibial tuberosity.
    E) head of the fibula.

    Answer: B

16) The "shinbone" is the
    A) femur.
    B) fibula.
    C) pubis.
    D) radius.
    E) tibia.

    Answer: E

17) The part of the tibia that articulates with the head of the fibula is the
    A) tibial tuberosity.
    B) medial malleolus.
    C) medial condyle.
    D) lateral condyle.
    E) None of these horizontal tibia does not articulate with the head of the femur.

    Answer: D

18) When someone says an elderly person "broke her hip," it is most likely the fracture occurred in the
    A) sacroiliac joint.
    B) pubic symphysis.
    C) neck of the femur.
    D) acetabulum.
    E) ilium.

    Answer: C

19) The prominence that can be felt on the medial surface of the ankle is part of the
    A) tibia.
    B) fibula.
    C) talus.
    D) calcaneus.
    E) navicular.

    Answer: A

20) The greater and lesser trochanters are projections seen on the
    A) humerus.
    B) scapula.
    C) tibia.
    D) femur.
    E) ischium.
    Answer: D

21) The distal end of the fibula articulates with the
    A) lateral condyle of the tibia.
    B) lateral condyle of the femur.
    C) fibular notch of the tibia.
    D) medial malleolus of the tibia.
    E) calcaneus.
    Answer: C

22) The lateral malleolus is part of the
    A) tibia.
    B) fibula.
    C) femur.
    D) humerus.
    E) ulna.
    Answer: B

23) During walking, the talus transmits about half of the weight of the body to the
    A) navicular.
    B) cuboid.
    C) calcaneus.
    D) metatarsals.
    E) phalanges.
    Answer: C

24) The head of metatarsal I articulates with the
    A) base of the proximal phalanx of the hallux.
    B) first cuneiform.
    C) cuboid.
    D) base of metatarsal II.
    E) hamate.
    Answer: A

25) The acromion process of the scapula articulates with the
    A) greater tubercle of the humerus.
    B) head of the humerus.
    C) lateral end of the clavicle.
    D) medial end of the clavicle.
    E) vertebral column.

    Answer: C

26) Which of the following is located on the ventral surface of the scapula?
    A) acromion process
    B) spine
    C) supraspinous fossa
    D) subscapular fossa
    E) All of these are correct.

    Answer: D

27) The medial part of the longitudinal arch includes ALL of the following EXCEPT the
    A) cuboid.
    B) cuneiforms.
    C) talus.
    D) navicular.
    E) heads of the three metatarsals I, II, and III.

    Answer: A

28) The roughened area on the middle portion of the shaft of the humerus is the
    A) deltoid tuberosity.
    B) anatomical neck.
    C) capitulum.
    D) trochlea.
    E) lesser tubercle.

    Answer: A

29) A lateral deviation of the proximal phalanx of the great toe and medial displacement of metatarsal I describes
    A) clawfoot.
    B) flatfoot.
    C) clubfoot.
    D) bunion.
    E) genu valgus.

    Answer: D

30) The prominence of the elbow is formed by the
   A) coronoid process of the ulna.
   B) olecranon process of the ulna.
   C) head of the radius.
   D) head of the humerus.
   E) ulnar tuberosity.

   Answer: B

31) The main function of the appendicular skeleton is to
   A) facilitate movement.
   B) protect internal organs.
   C) produce hormones for regulation of calcium balance.
   D) store iron for blood cell production.
   E) help regulate body temperature.

   Answer: A

32) Which of the following lists the proximal row of carpal bones in the correct order from lateral to medial?
   A) scaphoid, pisiform, lunate, triquetrum
   B) pisiform, triquetrum, lunate, scaphoid
   C) scaphoid, lunate, triquetrum, pisiform
   D) scaphoid, triquetrum, lunate, pisiform
   E) None of these, because these are the distal carpal bones

   Answer: C

33) Which of the following lists the distal row of carpal bones in the correct order from lateral to medial?
   A) hamate, capitate, trapezoid, trapezium
   B) trapezium, trapezoid, capitate, hamate
   C) capitate, hamate, trapezium, trapezoid
   D) trapezoid, trapezium, capitate, hamate
   E) None of these, because these are the proximal carpal bones

   Answer: E

34) The costal tuberosity is located on the
   A) lateral midshaft of the humerus.
   B) anterior surface of the distal end of the humerus.
   C) superior ramus of the pubis.
   D) posterior aspect of the femur.
   E) inferior surface of the medial end of the clavicle.

   Answer: E

35) ALL of the following are TRUE for the scapula EXCEPT the
    A) scapular notch is in the superior border.
    B) acromion is the high point of the shoulder.
    C) glenoid cavity opens toward the lateral side of the body.
    D) medial border articulates with the vertebral column.
    E) spine is on the posterior surface.

    Answer: D

36) The bones making up the palm of the hand are the
    A) metatarsals.
    B) metacarpals.
    C) carpals.
    D) tarsals.
    E) phalanges.

    Answer: B

37) The trochlea is located
    A) between the greater and lesser tubercles of the humerus.
    B) between the coronoid process and the olecranon of the ulna.
    C) on the lateral midshaft of the humerus.
    D) lateral to the capitulum of the humerus.
    E) medial to the capitulum of the humerus.

    Answer: E

38) Which metacarpal is proximal to the little finger?
    A) number I
    B) number II
    C) number III
    D) number IV
    E) number V

    Answer: E

39) When the forearm is flexed, which of the following is TRUE?
    A) The olecranon moves into the olecranon fossa.
    B) The coronoid process moves into the coronoid fossa.
    C) The radial head moves into the radial fossa.
    D) The radial head moves into the glenoid fossa.
    E) Both B and C are correct.

    Answer: B

40) When you sit on a stool, which part of the coxal bones touch the stool first?
    A) ischial spines
    B) ischial tuberosities
    C) iliac crests
    D) pubic symphysis
    E) inferior pubic rami

    Answer: B

41) The interosseous membrane joins the
    A) radius and ulna.
    B) bones of the metacarpals.
    C) femur and tibia.
    D) two scapulae.
    E) medial malleolus and lateral malleolus.

    Answer: A

42) The radius articulates with the
    A) lateral metacarpals.
    B) trapezoid.
    C) capitate.
    D) lunate.
    E) medial metacarpals.

    Answer: D

43) The superior border of the most superior of the subdivisions of the hipbone is the
    A) ischial tuberosity.
    B) superior ramus of the pubis.
    C) pubic crest.
    D) ischial spine.
    E) iliac crest.

    Answer: E

44) The head of the capitate articulates with the
    A) radius.
    B) head of metacarpal III.
    C) scaphoid.
    D) pisiform.
    E) hamate.

    Answer: E

45) The large hole in the coxal bone through which blood vessels and nerves pass is the
   A) acetabulum.
   B) pubic symphysis.
   C) obturator foramen.
   D) iliac fossa.
   E) glenoid cavity.

   Answer: C

46) The lesser sciatic notch is a feature of the
   A) ilium.
   B) ischium.
   C) pubis.
   D) femur.
   E) sacrum.

   Answer: B

47) The posterior landmark of the pelvic brim is the
   A) sacral promontory.
   B) posterior superior iliac spines.
   C) posterior inferior iliac spines.
   D) ischial tuberosities.
   E) tip of the coccyx.

   Answer: A

48) ALL of the following are TRUE for the true pelvis EXCEPT
   A) it surrounds the pelvic cavity.
   B) its inferior opening is the pelvic outlet.
   C) it is bounded anteriorly by the abdominal wall.
   D) it is bounded posteriorly by the sacrum and coccyx.
   E) it is the part of the bony pelvic below the pelvic brim.

   Answer: C

49) Which of the following is TRUE for the false pelvis?
   A) It does not normally contain pelvic organs.
   B) It is bounded posteriorly by the sacrum and coccyx.
   C) Its superior opening is the pelvic inlet.
   D) Its inferior opening is the pelvic outlet.
   E) It is bordered anteriorly by the pubic symphysis.

   Answer: A

50) A pair of coxal bones unearthed from an unmarked grave have oval-shaped obturator foramina and a pubic arch of greater than 90 degrees. What can you tell from this information?

    A) The person was probably very old because the bones have spread apart from the original position.

    B) The bones are probably those of a male because the pubic angle would be less in a female.

    C) The bones are probably those of a female because the pubic angle would be less in a male.

    D) The bones are probably from someone who suffered from a vitamin D deficiency, which caused abnormal flexibility in the bones.

    E) The bones are probably from a very young person because the obturator foramen is round in adults.

Answer: C

MATCHING. Choose the item in column 2 that best matches each item in column 1.

51)	Head of femur.	A.	Head of fibula
52)	Medial extremity of clavical.	B.	Acetabulum
53)	Lateral condyle of tibia.	C.	Cuneiforms
54)	Lateral malleolus of fibula.	D.	Talus
55)	Metatarsals.	E.	Sternum

Answers: 51) B. 52) E. 53) A. 54) D. 55) C.

56)	Head of humerus.	A.	Capitulum of humerus
57)	Medial condyle of tibia.	B.	Scaphoid
58)	Lateral extremity of clavicle.	C.	Medial condyle of femur
59)	Head of radius.	D.	Glenoid cavity
60)	Distal end of radius.	E.	Aromion of scapula

Answers: 56) D. 57) C. 58) E. 59) A. 60) B.

61)	Lesser sciatic notch.	A.	Clavicle
62)	Conoid tubercle.	B.	Humerus
63)	Trochlear notch.	C.	Ulna
64)	Lateral malleolus.	D.	Ischium
65)	Coronoid fossa.	E.	Fibula

Answers: 61) D. 62) A. 63) C. 64) E. 65) B.

66)	Fibular notch.	A.	Ilium
67)	Glenoid cavity.	B.	Scapula
68)	Ulnar notch.	C.	Radius
69)	Greater sciatic notch.	D.	Tibia
70)	Gluteal tuberosity.	E.	Femur

Answers: 66) D. 67) B. 68) C. 69) A. 70) E.

SHORT ANSWER. Write the word or phrase that best completes each statement or answers the question.

71) The scapula articulates with the clavicle and the _____.

Answer: humerus

72) The former site of the epiphyseal plate on the proximal end of the humerus is the _____.

Answer: anatomical neck

73) The fossa that is formed by the union of the ilium, ischium, and pubis, and that receives the head of the femur is the _____.

Answer: acetabulum

74) The medial end of the clavicle is the _____ extremity.

Answer: sternal

75) At the lateral end of the superior border of the scapula is a projection of the anterior surface called the _____ process.

Answer: coracoid

76) The only foot bone that articulates with the tibia and fibula is the _____.

Answer: talus

77) The head of the humerus articulates with the _____ of the _____.

Answer: glenoid cavity; scapula

78) The lateral end of the clavicle articulates with the _____ of the _____.

Answer: acromion; scapula

79) The spool-shaped surface at the distal end of the humerus that articulates with the ulna is the _____.

Answer: trochlea

80) The proximal end of the fibula is called the _____.

Answer: head

81) The distal end of the humerus has two indentations that receive parts of the ulna during flexion and extension of the forearm; these are the _____ fossa on the anterior surface and the _____ fossa on the posterior surface.

Answer: coronoid; olecranon

82) The ulna and radius are connected by a broad, flat fibrous connective tissue called the _____.

Answer: interosseous membrane

83) The pointed projection on the posterior side of the distal end of the ulna is the _____.
Answer: styloid process

84) The radial notch of the ulna and the capitulum of the humerus both articulate with the _____ of the _____.
Answer: head; radius

85) The head of the ulna articulates with the ulnar notch of the _____.
Answer: radius

86) The force of a fall on an outstretched hand is transmitted from the capitate to the radius through the _____.
Answer: scaphoid

87) The distal end of each metacarpal bone is its _____.
Answer: head

88) The most medial bone of the distal row of carpals is the _____.
Answer: hamate

89) The hip bones are also called _____ bones.
Answer: coxal

90) The three bones making up a single hip bone are the _____, the _____, and the _____.
Answer: ilium; ischium; pubis

91) The most laterally palpable bony landmark of the shoulder region is the _____ of the _____.
Answer: greater tubercle; humerus

92) The roughened, V-shaped area in the middle portion of the shaft of the humerus is the _____.
Answer: deltoid tuberosity

93) The sciatic nerve passes through a large indentation in the posterior part of the ilium called the _____.
Answer: greater sciatic notch

94) The anterior, posterior, and inferior gluteal lines are all on the lateral surface of the _____.
Answer: ilium

95) The beginning of the iliopectineal line is a projection called the _____.

Answer: pubic tubercle

ESSAY. Write your answer in the space provided or on a separate sheet of paper.

96) Describe the differences between the male pelvis and the female pelvis.

Answer: The female pelvis is specialized for childbirth in the following ways: false pelvis is shallower; pelvic brim is larger and more oval (vs. heart-shaped in male); pubic arch is greater than 90 degrees (male less than 90 degrees); ilium less vertical; iliac fossa shallower; iliac crest less curved; acetabulum smaller; obturator foramen oval (vs. rounded in male).

97) On her way to her car after class, Louise slipped on the ice and fell on her outstretched hand. She cried out, "I think I've broken something!" What are the weak points in the upper extremity and shoulder girdle that might be broken? Why are these places so vulnerable?

Answer: Weak points include: midregion of clavicle, distal end of radius, and scaphoid. These are the regions through which the mechanical force of the landing will be transmitted.

98) Describe the points of articulation among the proximal ends of the ulna and radius and the distal end of the humerus.

Answer: The following articulations are seen: trochlea of humerus fits into trochlear notch of ulna; head of radius articulates with capitulum and radial fossa of humerus; coronoid process of ulna fits into coronoid fossa of humerus during flexion of forearm; olecranon of ulna fits into olecranon fossa of humerus during extension of forearm; head of radius articulates with radial notch of ulna.

99) Carolyn has had a long career as a dental hygienist. After 15 years of making small, repetitive wrist and finger movements, she has been diagnosed with carpal tunnel syndrome. What is the "carpal tunnel?"

Answer: The carpal tunnel is a concavity formed by the pisiform, hamate, scaphoid, and trapezium, plus the flexor retinaculum, through which the flexor tendons of the digits and thumb and median nerve pass.

100) Describe/define the pelvic brim.

Answer: The pelvic brim is the boundary between the true and false pelvis, and is the circumference of an oblique plane formed as follows: begin posteriorly at sacral promontory, trace laterally and inferiorly along arcuate lines of pubis, anteriorly to superior portion of pubic symphysis.

CHAPTER 9   Joints

MULTIPLE CHOICE. Choose the one alternative that best completes the statement or answers the question.

1)   A synarthrosis is a(n)
   A) slightly movable joint.
   B) joint in which bones are connected by fibrocartilage.
   C) freely movable joint.
   D) joint with lubricating fluid between articulating bones.
   E) immovable joint.

   Answer: E

2)   Which of the following is a freely movable joint?
   A) amphiarthrosis
   B) synostosis
   C) diarthrosis
   D) synarthrosis
   E) symphysis

   Answer: C

3)   A movement that increases the angle between articulating bones is
   A) abduction.
   B) adduction.
   C) flexion.
   D) extension.
   E) rotation.

   Answer: D

4)   Which of the following best describes inversion?
   A) raising up on balls of feet to stand on toes
   B) rocking back on heels
   C) turning soles of feet away from each other
   D) turning soles of feet to face each other
   E) bending the knee to raise sole of foot toward back

   Answer: D

5) An amphiarthrosis is a(n)
   A) slightly movable joint.
   B) joint in which bones are connected by fibrocartilage.
   C) freely movable joint.
   D) joint with lubricating fluid between articulating bones.
   E) immovable joint.

   Answer: A

6) In a symphysis, articulating bones are
   A) connected by hyaline cartilage.
   B) connected by fibrocartilage.
   C) separated by a joint cavity.
   D) connected by a fibrous membrane.
   E) fused by new bone.

   Answer: B

7) Sutures and synchondroses are similar in that they both
   A) are functionally classified as amphiarthroses.
   B) have joint cavities between articulating bones.
   C) become synostoses as bodies age.
   D) are functionally classified as synarthroses.
   E) Both B and C are correct.

   Answer: E

8) A fibrocartilage disc that extends into a joint cavity describes a(n)
   A) ligament.
   B) bursa.
   C) meniscus.
   D) articular cartilage.
   E) synovial membrane.

   Answer: C

9) A sac of synovial fluid between bones and overlying tissues describes a(n)
   A) ligament.
   B) bursa.
   C) meniscus.
   D) syndesmosis.
   E) gomphosis.

   Answer: B

10) Dense connective tissue connecting one bone to another bone describes a(n)

A) ligament.

B) bursa.

C) meniscus.

D) tendon.

E) synovial membrane.

Answer: A

11) Hyaline cartilage makes up the

A) articular discs.

B) connection between bones in sutures.

C) bursae.

D) articular cartilage.

E) All of the above are correct.

Answer: D

12) Synovial fluid is produced by

A) chondrocytes in the articular cartilage.

B) osteocytes in the articulating bones.

C) cells in the inner layer of the articular capsule.

D) fibroblasts in ligaments surrounding the joint.

E) chondrocytes in menisci.

Answer: C

13) Which of the following is functionally classified as an amphiarthrosis?

A) suture

B) ball-and-socket joint

C) syndesmosis

D) planar joint

E) Both A and B are correct.

Answer: C

14) Flexion and extension are the principal movements performed at

A) planar joints.

B) pivot joints.

C) syndesmoses.

D) symphyses.

E) hinge joints.

Answer: E

15) The greatest range of motion occurs at
    A) hinge joints.
    B) ellipsoidal joints.
    C) pivot joints.
    D) ball-and-socket joints.
    E) synchondroses.

    Answer: D

16) The fibrous connective tissue that connects a muscle to a bone is called a(n)
    A) tendon.
    B) meniscus.
    C) ligament.
    D) bursa.
    E) labrum.

    Answer: A

17) The type of movement normally seen at pivot joints is
    A) abduction and adduction.
    B) rotation.
    C) flexion and extension.
    D) protraction and retraction.
    E) inversion and eversion.

    Answer: B

18) Which of the following is an example of a symphysis?
    A) coxal joint
    B) intercarpal joint
    C) intervertebral joint
    D) joint between frontal and parietal bones
    E) Both A and B are correct.

    Answer: C

19) Which of the following is an example of a planar joint?
    A) coxal joint
    B) intercarpal joint
    C) intervertebral joint
    D) joint between frontal and parietal bones
    E) Both A and B are correct.

    Answer: B

20) Doing a backbend is an example of

   A) flexion.

   B) hyperflexion.

   C) hyperextension.

   D) rotation.

   E) dorsiflexion.

   Answer: C

21) Which of the following is an example of a ball-and-socket joint?

   A) tibiofemoral joint

   B) glenohumeral joint

   C) atlanto-axial joint

   D) temporomandibular joint

   E) Both B and D are correct.

   Answer: B

22) The joint formed by the root of a tooth and its socket is called a

   A) condyloid joint.

   B) syndesmosis.

   C) synchondrosis.

   D) gomphosis.

   E) synostosis.

   Answer: D

23) The dense connective tissue joining structure in a gomphosis is called a(n)

   A) articular disc.

   B) bursa.

   C) tendon.

   D) synovial membrane.

   E) periodontal ligament.

   Answer: E

24) Which of the following is an example of a synchondrosis?

   A) intervertebral joint

   B) epiphyseal plate

   C) intertarsal joint

   D) joint between root of tooth and alveolus

   E) Both A and B are correct.

   Answer: B

25) Turning the palm posterior or inferiorly is referred to as

   A) dorsiflexion of the hand.

   B) abduction of the hand

   C) supination of the forearm.

   D) pronation of the forearm.

   E) rotation of the hand.

   Answer: D

26) The mouth is opened when

   A) depression occurs at the temporomandibular joint.

   B) elevation occurs at the temporomandibular joint.

   C) flexion occurs at the temporomandibular joint.

   D) extension occurs at the temporomandibular joint.

   E) protraction occurs at the temporomandibular joint.

   Answer: A

27) Standing on the toes, such as ballet dancers do, is an example of

   A) inversion.

   B) eversion.

   C) plantar flexion.

   D) dorsiflexion.

   E) hyperextension.

   Answer: C

28) Making circles with the arms is best described as

   A) abduction.

   B) hyperextension.

   C) circumduction.

   D) elevation.

   E) flexion.

   Answer: C

29) The transverse humeral ligament extends from the greater tubercle of the humerus to the

   A) anatomical neck of the humerus.

   B) coracoid process of the scapula.

   C) glenoid labrum.

   D) lesser tubercle of the humerus.

   E) deltoid tuberosity of the humerus.

   Answer: D

30) Fibrocartilage that stabilizes the shoulder joint is the
   A) articular capsule.
   B) coracohumeral ligament.
   C) glenoid labrum.
   D) subscapular bursa.
   E) Both A and C are fibrocartilage structures.

   Answer: C

31) Spreading the legs apart, as in doing jumping jacks, is an example of
   A) abduction.
   B) adduction.
   C) flexion.
   D) extension.
   E) protraction.

   Answer: A

32) Articular fat pads are accumulations of adipose tissue in the
   A) articular cartilage.
   B) articular discs.
   C) synovial membrane.
   D) bursae.
   E) extracapsular ligaments.

   Answer: C

33) The anterior cruciate ligament extends from the lateral condyle of the femur to the
   A) medial condyle of the femur.
   B) lateral epicondyle of the femur.
   C) medial condyle of the femur.
   D) head of the fibula.
   E) area anterior to the intercondylar eminence of the tibia.

   Answer: E

34) The patellar ligament extends from the patella to the
   A) linea aspera of the femur.
   B) tibial tuberosity of the tibia.
   C) head of the fibula.
   D) medial and lateral condyles of the femur.
   E) lesser trochanter of the femur.

   Answer: B

35) The synovial membrane is made of
A) areolar connective tissue.
B) mesothelium.
C) hyaline cartilage.
D) dense irregular connective tissue.
E) endothelium.

Answer: A

36) Synovial fluid consists of interstitial fluid and
A) phospholipids.
B) hydroxyapatite.
C) collagen.
D) hyaluronic acid.
E) elastin.

Answer: D

37) Subdivision of the synovial cavity and directing the flow of synovial fluid are functions of
A) articular cartilage.
B) bursae.
C) menisci.
D) extracapsular ligaments.
E) Both B and C are correct.

Answer: C

38) The type of movement possible at a planar joint is
A) gliding.
B) flexion.
C) circumduction.
D) abduction.
E) All of the above are possible.

Answer: A

39) Bursae are filled with
A) venous blood.
B) air.
C) synovial fluid.
D) dense irregular connective tissue.
E) adipose tissue.

Answer: C

40) You can elevate and depress your
   A) forearms.
   B) legs.
   C) mandible.
   D) feet at the ankles.
   E) All of these can be elevated and depressed.

   Answer: C

41) When your palms are lying flat on the table, your forearms are
   A) inverted.
   B) everted.
   C) pronated.
   D) supinated.
   E) dorsiflexed.

   Answer: C

42) You are bent over at the waist touching your toes. ALL of the following body parts are extended EXCEPT your
   A) fingers.
   B) legs.
   C) arms.
   D) vertebral column.
   E) elbows.

   Answer: D

43) You are sitting up straight in a chair with your fingers curled into your palm on the table in front of you. ALL of the following body parts are flexed EXCEPT your
   A) fingers.
   B) forearms.
   C) legs.
   D) knees.
   E) vertebral column.

   Answer: E

44) ALL of the following are normal movements of the shoulder EXCEPT
   A) abduction.
   B) hyperextension.
   C) circumduction.
   D) lateral rotation.
   E) medial rotation.

   Answer: B

45) The coracohumeral ligament extends from the
   A) conoid tubercle of the clavicle to the greater tubercle of the humerus.
   B) acromion of the scapula to the greater tubercle of the humerus.
   C) coracoid process of the scapula to the greater tubercle of the humerus.
   D) coracoid process of the scapula to the anatomical neck of the humerus.
   E) coracoid process of the scapula to the lesser tubercle of the humerus.

   Answer: C

46) The medial epicondyle of the humerus and the coronoid process of the ulna are attachment points for the
   A) transverse humeral ligament.
   B) subcoracoid bursa.
   C) radial collateral ligament.
   D) ulnar collateral ligament.
   E) coracohumeral ligament.

   Answer: D

47) The acetabular labrum is
   A) an accessory ligament attaching the fovea capitis of the femur to the acetabulum.
   B) a ring of circular collagen fibers within the articular capsule of the hip that surrounds the neck of the femur.
   C) an accessory ligament running from the acetabulum to the intertrochanteric line of the femur.
   D) the central, deepest point of the acetabulum to which the fovea capitis of the femur connects.
   E) the fibrocartilage rim of the acetabulum that enhances its depth.

   Answer: E

48) Abduction and adduction are possible for ALL of the following EXCEPT the
   A) coxal joint.
   B) glenohumeral joint.
   C) metacarpophalangeal joint.
   D) radiocarpal joint.
   E) elbow.

   Answer: E

49) Rotation is the only movement possible at the
   A) radioulnar joint.
   B) radiocarpal joint.
   C) vertebrocostal joints.
   D) atlanto-occipital joint.
   E) temporomandibular joint.

   Answer: A

50) Which of the following is considered to be an autoimmune disease?

    A) rheumatoid arthritis
    B) osteoarthritis
    C) gouty arthritis
    D) osteosarcoma
    E) Both A and C are autoimmune diseases.

    Answer: A

MATCHING. Choose the item in column 2 that best matches each item in column 1.

51)	Movement of the mandible downward.	A.	Depression
52)	Bending of the foot in the direction of the upper surface.	B.	Dorsiflexion
53)	Movement of the forearm in which the palm is turned anteriorly or superiorly.	C.	Eversion
54)	Movement of the sole of the foot outward so that the soles face away from each other.	D.	Supination
55)	Movement of the mandible forward on the plane parallel to the ground.	E.	Protraction

Answers: 51) A. 52) B. 53) D. 54) C. 55) E.

56)	Bending of the foot in the direction of the sole.	A.	Plantar flexion
57)	Movement of the mandible backward on a plane parallel to the ground.	B.	Retraction
58)	Movement of the mandible upward.	C.	Elevation
59)	Movement of the forearm in which the palm is turned posteriorly or inferiorly.	D.	Pronation
60)	Movement of the sole of the foot outward so that the soles face toward each other.	E.	Inversion

Answers: 56) A. 57) B. 58) C. 59) D. 60) E.

61)	Roots of teeth in alveoli.	A.	Gomphosis
62)	Interphalangeal joint.	B.	Hinge joint
63)	Joint between frontal and parietal bones.	C.	Suture
64)	Intertarsal joint.	D.	Plantar joint
65)	Atlanto-axial joint.	E.	Pivot joint

Answers: 61) A. 62) B. 63) C. 64) D. 65) E.

66) Intervertebral joint.	A. Symphysis
67) Distal tibiofibular joint.	B. Syndesmosis
68) Eiphyseal plate.	C. Synchondrosis
69) Glenohumeral joint.	D. Ball-and-socket joint
70) Radiocarpal joint.	E. Condyloid joint

Answers: 66) A. 67) B. 68) C. 69) D. 70) E.

SHORT ANSWER. Write the word or phrase that best completes each statement or answers the question.

71) The angle between articulating bones is decreased by a movement called _____.

Answer: flexion

72) A bone moves away from the body's midline during _____.

Answer: abduction

73) The atlanto-axial joint is an example of a freely movable joint called a(n) _____.

Answer: pivot

74) The trapezium and metacarpal I form a freely movable joint called a(n) _____ joint.

Answer: saddle

75) A partial or incomplete dislocation is called a(n) _____.

Answer: subluxation

76) A joint in which more flexible connective tissue has been replaced by bone as the body ages is called a(n) _____.

Answer: synostosis

77) The interosseous membrane between the tibia and fibula is an example of a(n) _____.

Answer: syndesmosis

78) In a symphysis, the articulating bones are joined by _____.

Answer: fibrocartilage

79) The dense connective tissue holding a tooth in its socket is called the _____.

Answer: periodontal ligament

80) All synovial joints are classified functionally as _____.

Answer: diarthroses

81) Fibrous connective tissue that connects one bone to another in a joint capsule is called a(n) _____.
Answer: ligament

82) Nutrients are supplied to the chondrocytes of the articular cartilage by _____.
Answer: synovial fluid

83) Fluid-filled sacs located between bones and overlying tissues that help alleviate pressure are called _____.
Answer: bursae

84) Someone who has a "torn cartilage" in the knee has damaged a(n) _____.
Answer: articular disc (meniscus)

85) Examination of the interior of a joint is known as _____.
Answer: arthroscopy

86) Intercarpal joints are examples of synovial joints known as _____ joints.
Answer: planar

87) Bending the trunk backward at the intervertebral joints is an example of _____.
Answer: hyperextension

88) The combined movements of flexion, extension, abduction, and adduction is called _____.
Answer: circumduction

89) A hormone that increases the flexibility of the fibrocartilage of the pubic symphysis is _____.
Answer: relaxin

90) Pain in a joint is called _____.
Answer: arthralgia

91) The glenohumeral joint and the coxal joint are the only examples of the _____ joint.
Answer: ball-and-socket

92) Movement of the thumb across the palm to touch the fingertips on the same hand is called _____.
Answer: opposition

93) Movement of the trunk in the frontal plane is called _____.
Answer: lateral flexion

94) The parietal and occipital bones are held together by a joint called a(n) _____.

Answer: suture

95) The tendons of the supraspinatus, infraspinatus, teres minor, and subscapularis muscles form a structure called the _____.

Answer: rotator cuff

ESSAY. Write your answer in the space provided or on a separate sheet of paper.

96) Describe the factors affecting the range of motion at synovial joints.

Answer:
1. Structure or shape of articulating bones—i.e., how closely they fit together
2. Strength and tension of ligaments—restricts direction and degree of movement
3. Arrangement and tension of muscles—complements ligaments
4. Apposition of soft parts—i.e., point at which one body surface meets another during a movement
5. Hormones—may affect flexibility of connective tissues
6. Disuse—decreases synovial fluid, connective tissue flexibility and muscle mass

97) Describe the affects of aging on joints.

Answer: Changes include: decreased production of synovial fluid; thinning of articular cartilage; ligaments shorten and lose flexibility; osteoarthritis develops from wear and tear—bone exposed at joints, spurs form and synovial membrane becomes inflamed, all affecting range of motion.

98) Describe the functions of synovial fluid.

Answer:
1. lubrication of joints
2. shock absorption
3. providing nutrients to articular cartilage
4. removal of debris and microbes via phagocyte activity
5. removal of metabolic wastes from articular cartilage

99) Name and briefly describe the common types of arthritis.

Answer:
1. Rheumatoid arthritis is an autoimmune disease in which the immune system attacks cartilage and joint linings, causing swelling, pain, and loss of function. Bones may fuse, making them immovable.
2. Osteoarthritis results from deterioration of articular cartilage due to wear and tear. Bone spurs form, which restricts movement.
3. Gouty arthritis occurs due to deposition of uric acid crystals in soft tissues of joints, eroding cartilage and causing inflammation.

100) Roger has spent 20 years pitching fastballs during his childhood and adult career. An unfortunate automobile accident dislocated his shoulder. Although he received immediate medical treatment, he can still feel his shoulder begin to slip out of joint every time he tries to raise his arms to begin his pitching motion. What could be the anatomical problem?

Answer: A torn rotator cuff weakens support that holds the head of the humerus in the glenoid cavity. A torn glenoid labrum makes the glenoid cavity shallower. Both allow the large humeral head to slip more easily out of relatively shallow glenoid cavity.

CHAPTER 10   Muscle Tissue

MULTIPLE CHOICE. Choose the one alternative that best completes the statement or answers the question.

1) Striated (skeletal) muscle tissue
   A) is composed of long, spindle-like cells, each containing a single nucleus.
   B) has the ability to contract when stimulated.
   C) is present in the walls of arteries.
   D) has the ability to contract rhythmically by itself.
   E) is described by all of the above.

   Answer: B

2) Which of these is true?
   A) Most of the muscle tissue of the body is smooth (nonstriated) muscle.
   B) Skeletal muscle fibers lack voluntary control.
   C) Skeletal muscle fibers are under voluntary control.
   D) Striated muscles are those without antagonists.
   E) None of the above.

   Answer: C

3) If a tissue has striations and several flattened peripheral nuclei per fiber, then that tissue could be
   A) smooth muscle.
   B) cardiac muscle.
   C) visceral muscle.
   D) skeletal muscle.

   Answer: D

4) Cylindrical muscle cells that contain multiple nuclei located peripherally within the cell would be
   A) skeletal muscle cells only.
   B) single-unit smooth muscle cells only.
   C) multiunit smooth muscle cells only.
   D) cardiac muscle cells.
   E) Both single-unit and multiunit smooth muscle cells.

   Answer: A

5) The ability of a muscle to stretch without being damaged is
   A) excitability.
   B) contractility.
   C) extensibility.
   D) elasticity.
   E) conductivity.

   Answer: C

6) A tendon is composed of
   A) perimysium.
   B) epimysium.
   C) endomysium.
   D) All are correct.

   Answer: D

7) The endomysium is
   A) the areolar connective tissue of the hypodermis.
   B) dense irregular connective tissue holding muscles with similar functions together.
   C) areolar connective tissue surrounding individual muscle fibers.
   D) dense regular connective tissue attaching muscle to bone.
   E) the site of calcium ion storage in skeletal muscle.

   Answer: C

8) The sarcolemma is the
   A) storage site for calcium ions in myofibers.
   B) plasma membrane of a myofiber.
   C) compound that binds oxygen for use in slow, oxidative muscle cells.
   D) separation between sarcomeres in a myofiber.
   E) structure that produces acetylcholine.

   Answer: B

9) Muscle atrophy is the result of loss of
   A) actin.
   B) myosin.
   C) mitochondria.
   D) sarcoplasmic reticulum.
   E) All are correct.

   Answer: E

10) Myosin is seen in
   A) thin filaments.
   B) thick filaments.
   C) the sarcoplasmic reticulum.
   D) the sarcolemma.
   E) Both A and B are correct.

   Answer: B

11) Skeletal muscle fibers are multinucleate because
   A) the original nucleus reproduced as the cell grew.
   B) when neighboring cells undergo apoptosis, cells that remain absorb their nuclei before they degenerate.
   C) the fibers grew so fast they engulfed any cells in their way.
   D) the fibers formed from the fusion of many smaller cells during embryonic development.
   E) All of the above have occurred.

   Answer: D

12) Thin myofilaments of a striated muscle fiber are anchored to the
   A) Z-lines.
   B) transverse tubules.
   C) H-zones.
   D) terminal cisternae.
   E) myosin.

   Answer: A

13) Glucose is stored in muscle cells as
   A) actin.
   B) myoglobin.
   C) glycogen.
   D) creatine.
   E) calmodulin.

   Answer: C

14) The function of calcium ions in skeletal muscle contraction is to
    A) bind to receptors on the sarcolemma at the neuromuscular junction to stimulate muscle contraction.
    B) cause a pH change in the sarcoplasm to trigger muscle contraction.
    C) bind to the myosin binding sites on actin so that myosin will have something to attach to.
    D) bind to the troponin on the thin filaments so that the myosin binding sites on the actin can be exposed.
    E) bind oxygen to fuel cellular respiration in oxidative fibers.

    Answer: D

15) Skeletal muscles are stimulated to contract when
    A) calcium ions bind to the sarcolemma, causing an action potential.
    B) acetylcholine diffuse into the sarcoplasm.
    C) acetylcholine binds to receptors on the sarcolemma, causing an action potential.
    D) ATP is released from the sarcoplasmic reticulum.
    E) oxygen binds to myoglobin.

    Answer: C

16) During the sliding filament mechanism of muscle contraction, ATP attaches to
    A) actin.
    B) troponin.
    C) myosin.
    D) tropomyosin.
    E) dystrophin.

    Answer: C

17) What structures meet at the neuromuscular junction?
    A) T tubules and sarcoplasmic reticulum
    B) the sarcolemma and T tubules
    C) an axon and the sarcoplasmic reticulum
    D) an axon and the sarcolemma
    E) an axon and thick myofilaments

    Answer: D

18) The function of myoglobin is to
    A) bind oxygen for aerobic respiration.
    B) bind actin to shorten myofibrils.
    C) block the myosin binding sites on thin filaments.
    D) store ATP.
    E) separate one sarcomere from another.

    Answer: A

19) After death, cellular membranes leak _____ to cause muscle contraction or rigor mortis.
   A) calcium
   B) sodium
   C) potassium
   D) hydrogen
   E) All are correct.

   Answer: A

20) The Sliding Filament Theory of muscle contraction says that myofibers shorten when
   A) actin filaments become shorter when they combine with myosin heads.
   B) thin filaments are pulled toward the center of the sarcomere by swiveling of the myosin heads.
   C) myosin heads rotate when they attach to actin, causing the myosin filaments to fold in the middle.
   D) acetylcholine reduces the friction between thin and thick filaments, so they slide over each other more easily.
   E) a neurotransmitter alters the arrangement of collagen fibers in the endomysium, causing it to shrink more tightly around the myofiber.

   Answer: B

21) The refractory period is the time
   A) between stimulation and the start of contraction.
   B) when the muscle is shortening.
   C) when the muscle is lengthening.
   D) following a stimulus during which a muscle cell cannot respond to another stimulus.
   E) it takes for acetylcholine to cross the synaptic cleft.

   Answer: D

22) Fused tetanus is
   A) a pathological condition in which stimulation of one muscle triggers contraction of all others in the same motor unit.
   B) a sustained contraction with partial relaxation between stimuli.
   C) a sustained contraction in which individual twitches cannot be discerned.
   D) a brief contraction of all the fibers in a motor unit.
   E) a phenomenon that occurs in only muscle cells without refractory periods.

   Answer: C

23) The role of acetylcholine in skeletal muscle contractions is to
    A) diffuse into the sarcoplasm to open calcium ion channels in the sarcoplasmic reticulum.
    B) bind to troponin to expose myosin binding sites.
    C) hydrolyze ATP to release energy.
    D) bind to specific receptors on the sarcolemma to open sodium ion channels.
    E) react with titin molecules to cause relaxation of a muscle fiber.

    Answer: D

24) A motor unit is
    A) all the muscles that act as prime movers for a particular action.
    B) the sarcomeres in an individual myofibril.
    C) all of the neurons that stimulate a particular skeletal muscle.
    D) a motor neuron plus all the skeletal muscle fibers it stimulates.
    E) the quantity of neurotransmitter that is sufficient to stimulate muscle contraction.

    Answer: D

25) The effect of acetylcholine on skeletal muscle fibers ends when
    A) an action potential is generated.
    B) myosin attaches to actin.
    C) sodium ions begin diffusing out of the fiber.
    D) it is broken down by acetylcholinesterase.
    E) a phosphate attaches to creatine.

    Answer: D

26) The Z disc is the
    A) separation between sarcomeres.
    B) area of a sarcomere where only thick myofilaments are found.
    C) point where the axon and the sarcolemma meet.
    D) place where two cardiac muscle cells meet.
    E) electrical voltage at which a muscle cell fires an action potential.

    Answer: A

27) Muscle paralysis occurs when anticholinesterase drugs are released, this is
    A) botulinum toxin.
    B) curare.
    C) neostigmine.
    D) All are correct.

    Answer: C

28) The muscle protein whose function is related to its golf club-like shape is
   A) actin.
   B) troponin.
   C) tropomyosin.
   D) myosin.
   E) titin.

   Answer: D

29) The neurotransmitter molecule released at the neuromuscular junction is known as
   A) calmodulin.
   B) myasthenia.
   C) acetylcholine.
   D) phosphagen.
   E) All are correct.

   Answer: C

30) Which of the following conditions exists in a resting myofiber?
   A) Levels of calcium ions are relatively high in the sarcoplasm.
   B) Thin myofilaments extend across the H zone.
   C) ATP is attached to myosin cross bridges.
   D) Levels of acetylcholine are high inside the sarcoplasmic reticulum.
   E) Both A and C are correct.

   Answer: C

31) The function of dystrophin is to
   A) bind thick filaments to each other.
   B) hold thin filaments to the Z disc.
   C) form a latticework that supports all other muscle proteins.
   D) block the myosin binding sites on actin.
   E) link thin filaments to integral proteins of the sarcolemma.

   Answer: E

32) The protein responsible for much of the elasticity and extensibility of myofibrils is
   A) nebulin.
   B) alpha-actinin.
   C) titin.
   D) troponin.
   E) tropomyosin.

   Answer: C

33) The protein that helps reinforce the sarcolemma and transmit tension generated by the sarcomeres to the tendons is

A) actin.

B) nebulin.

C) myomesin.

D) dystrophin.

E) troponin.

Answer: D

34) The term *power stroke,* when used relative to skeletal muscle contraction, refers to the

A) flooding of the sarcoplasm with calcium ions.

B) release of acetylcholine from the motor neuron via exocytosis.

C) sudden increase in the sarcolemma's permeability to sodium ions following the binding of acetylcholine.

D) pulling of the tendon on the bone as a whole muscle contracts.

E) swiveling of the myosin heads as they combine with actin.

Answer: E

35) What is happening during the latent period of a muscle contraction?

A) Nothing.

B) Calcium ions are being actively transported back into the sarcoplasmic reticulum.

C) New contractile proteins are being synthesized.

D) Calcium ions are beginning to enter the sarcoplasm from the sarcoplasmic reticulum.

E) ATP molecules are attaching to myosin heads.

Answer: D

36) The compound that is stored as an energy reserve for skeletal muscle contraction is

A) phosphocreatine.

B) ADP.

C) magnesium phosphate.

D) calcium phosphate.

E) calmodulin.

Answer: A

37) Two molecules of pyruvic acid are the products of

A) reactions catalyzed by creatine kinase.

B) the breakdown of glucose via glycolysis.

C) aerobic cellular respiration.

D) the binding of actin and myosin.

E) removal of oxygen from myoglobin.

Answer: B

38) The fate of pyruvic acid formed during muscle activity depends on the

A) amount of oxygen available.

B) rate of creatine kinase activity.

C) relative proportions of actin and myosin present.

D) number of nuclei per cell.

E) surface area of the sarcolemma.

Answer: A

39) Most of the lactic acid remaining after exercise is

A) converted to glycogen and stored.

B) converted to creatine and stored.

C) converted to pyruvic acid and used for ATP production.

D) used as a building block for new contractile proteins.

E) stored as fat.

Answer: C

40) Slow oxidative fibers are said to be "slow" because they

A) break down acetylcholine slowly.

B) conduct action potentials slowly.

C) manufacture creatine phosphate slowly.

D) recover from fatigue slowly.

E) hydrolyze ATP slowly.

Answer: E

41) Muscle cells with relatively few mitochondria that generate most of their ATP via glycolysis and that have low resistance to fatigue are most likely

A) slow oxidative skeletal muscle fibers.

B) fast oxidative skeletal muscle fibers.

C) fast glycolytic skeletal muscle fibers.

D) cardiac muscle fibers.

E) smooth muscle fibers.

Answer: C

42) The phenomenon of oxygen debt is believed to be caused by the buildup of

A) actin.

B) lactic acid.

C) creatine phosphate.

D) ATP.

E) All are correct.

Answer: B

43) The role of tropomyosin in skeletal muscle is to
    A) provide an additional source of ATP during periods of high-intensity exercise.
    B) bind to myosin cross bridges during the power stroke.
    C) block the myosin binding sites on actin during periods of rest.
    D) actively transport calcium ions back into the sarcoplasmic reticulum following the power stroke.
    E) connect the thick myofilaments to the Z discs.

    Answer: C

44) When one stimulates during refractory period before the muscle has completely relaxed, the muscle increases in strength of contraction, these stimuli are called
    A) summation.
    B) muscle tone.
    C) isometric contraction.
    D) isotonic contraction.

    Answer: A

45) Cardiac muscle can stay contracted 10–15 times longer than skeletal muscle because
    A) intercalated discs are stronger than Z discs.
    B) there is less acetylcholinesterase available in cardiac muscle.
    C) channels allowing calcium ions to enter from the extracellular fluid add to that released from the sarcoplasmic reticulum.
    D) the contractile proteins in the sarcomeres are different.
    E) the passive tension on the heart is greater.

    Answer: C

46) The type of tissue in the iris that causes very precise adjustments in pupil diameter is
    A) visceral smooth muscle.
    B) multiunit smooth muscle.
    C) cardiac muscle.
    D) skeletal muscle.
    E) dense regular connective tissue.

    Answer: B

47) Which one of the following is NOT a characteristic of fast-twitch white fibers?
    A) contain much glycogen.
    B) geared to anaerobic metabolic processes.
    C) well supplied with blood vessels.
    D) do not contain large amounts of myoglobin.

    Answer: C

48) People with a higher proportion of fast glycolytic fibers often excel in activities that require periods of intense activity such as
   A) long-distance running.
   B) long-distance swimming.
   C) weight lifting.
   D) All are correct.

   Answer: C

49) An abundance of fast glycolytic fibers is seen in
   A) the walls of large arteries because they must stretch.
   B) neck muscles because they must stay contracted for long periods to hold up the head.
   C) arm muscles because they must move quickly and intensely.
   D) leg muscles because they must move constantly.
   E) the wall of the heart because it must maintain autorhythmicity.

   Answer: C

50) Anabolic steroids are similar to
   A) estrogens.
   B) progesterones.
   C) testosterones.
   D) cortisones.
   E) All are correct.

   Answer: C

MATCHING. Choose the item in column 2 that best matches each item in column 1.

	Column 1		Column 2
51)	Sequester calcium.	A)	Troponin
52)	Myosin binding site.	B)	Tropomyosin
53)	Fingers make up the thick filaments.	C)	Sarcoplasmic reticulum
54)	Calcium attaches during the contraction cycle.	D)	Actin
		E)	Myosin

Answers: 51) C. 52) D. 53) E. 54) A.

	Column 1		Column 2
55)	Neurotransmitter of neuromyojunction.	A)	Curare
56)	Blocks neuromyojunctions.	B)	Neostigmine
57)	Antiacetylcholinesterase inhibitor.	C)	Acetylcholine
58)	Blocks release of exocytosis of synaptic vesicles.	D)	Clostridium botulinum
		E)	Anabolic steroids

Answers: 55) C. 56) A. 57) B. 58) D.

59) Buildup of lactic acid.

60) Glucose storage form.

61) High-energy storage form.

62) Aerobic cellular respiration produces.

A) Creatine phosphate
B) ATP
C) Glycogen
D) Oxygen debt
E) Myoglobin

Answers: 59) D. 60) C. 61) A. 62) B.

63) Disease with damage to neuromyojunctions.

64) Disease with degeneration of skeletal muscle fibers.

65) Disorder affecting fibrous connective tissue.

66) Spasmodic contraction of skeletal muscle.

A) Fibromyalgia
B) Muscular dystrophy
C) Cramp
D) Myasthenia gravis
E) Tremor

Answers: 63) D. 64) B. 65) A. 66) C.

67) Increased muscle tone.

68) Pain associated with skeletal muscle.

69) A tumor consisting of muscle tissue.

70) Inflammation of muscle tissue.

A. Hypotonia
B. Hypertonia
C. Myoma
D. Myalgia
E. Myositis

Answers: 67) B. 68) D. 69) C. 70) E.

SHORT ANSWER. Write the word or phrase that best completes each statement or answers the question.

71) Muscle tissue that is both nonstriated and involuntary is _____.

Answer: smooth muscle tissue

72) The gap between a motor neuron and a muscle cell is called the _____.

Answer: synaptic cleft

73) The inability of a muscle to contract forcefully after prolonged activity is called _____.

Answer: muscle fatigue

74) The smallest, most fatigue-resistant skeletal muscle fibers are the _____ fibers.

Answer: slow oxidative

75) An oxygen-binding protein in skeletal muscle cells is _____.

Answer: myoglobin

76) The dense irregular connective tissue that carries nerves and blood vessels, fills the spaces between muscles, and separates muscles into functional groups is the _____.

Answer: deep fascia

77) The region of a sarcolemma adjacent to the axon terminals at a neuromuscular junction is called the _____.

Answer: motor end plate

78) The region of the sarcomere that contains only thick myofilaments is the _____.

Answer: H zone

79) The structures in smooth muscle fibers that are functionally analogous to the Z discs of skeletal muscle fibers are the _____.

Answer: dense bodies

80) A regulatory protein that binds calcium ions in the cytosol of smooth muscle fibers is _____.

Answer: calmodulin

81) The stem cells of smooth muscle cells are called _____.

Answer: pericytes

82) During embryonic development mesoderm destined to become skeletal muscle forms columns that undergo segmentation into a series of blocks of cells called _____.

Answer: somites

83) A disease that results from inappropriate production of antibodies that block acetylcholine receptors is _____.

Answer: myasthenia gravis

84) Myosin binding sites on actin are exposed when troponin changes shape as a result of binding _____.

Answer: calcium ions

85) Muscle contraction without muscle shortening is called a(n) _____ contraction.

Answer: isometric

86) The time between the application of a stimulus and the beginning of contraction, when calcium ions are being released from the sarcoplasmic reticulum, is called the _____ period.

Answer: latent

87) The time following a stimulus during which a muscle cell is unable to respond to another stimulus is called the _____ period.

Answer: refractory

88) An inherited myopathy that results in the rupture and death of muscle fibers is _____.

Answer: muscular dystrophy

89) Involuntary inactivation of a small number of motor units causes sustained, small contractions that give relaxed skeletal muscle a firmness known as _____.

Answer: muscle tone

90) The contractile protein in thin myofilaments is _____.

Answer: actin

91) The palest muscle fibers that are not fatigue-resistant and that get their ATP primarily via anaerobic respiration are _____ fibers.

Answer: fast glycolytic

92) Muscle fibers that are branched with a single, central nucleus are _____ muscle fibers.

Answer: cardiac

93) A muscle contraction in which the force developed by the muscle remains almost constant while the muscle shortens is said to be a(n) _____ contraction.

Answer: isotonic

94) An isotonic contraction in which the muscle lengthens to produce movement and to increase the angle at a joint is called a(n) _____ contraction.

Answer: eccentric

95) In the phosphagen system, high-energy phosphate groups can be stored for future ATP production by combining with _____.

Answer: creatine

ESSAY. Write your answer in the space provided or on a separate sheet of paper.

96) Outline the steps of skeletal muscle contraction.

Answer:
1. Stimulus provided by binding of ACh to the sarcolemma.
2. Resulting action potential travels along sarcolemma and into T tubules, triggering release of calcium ions from SR.
3. Calcium ions bind to troponin; resulting shape change causes myosin binding site to be exposed.
4. Myosin heads bind to actin, and swivel (power stroke), pulling Z discs closer together, shortening myofiber.
[add details of ATP use as desired]

97) Describe the mechanisms by which skeletal muscle tissue obtains ATP to fuel contraction.

Answer:
1. ATP is attached to resting myosin heads.
2. Creatine stores high-energy phosphate groups that can be added to ADP as needed.
3. Anaerobic glycolysis breaks down glucose into two molecules of pyruvic acid, releasing enough energy to net 2 ATP molecules per glucose.
4. Mitochondria oxidize pyruvic acid to carbon dioxide and water via aerobic cellular respiration. Energy released nets about 36 ATP molecules per glucose molecule, plus heat.

98) Compare and contrast the processes by which striated and nonstriated muscle tissues contract.

Answer: Smooth muscle: contraction begins more slowly; lasts longer due to slow influx of calcium ions; regulator protein (calmodulin) binds calcium ions and activates myosin light chain kinase to phosphorylate myosin and trigger attachment to actin; wide variety of stimuli.

Striated muscle: regulator protein (troponin) binds calcium ions, changes shape, and exposes myosin binding sites on actin, causing spontaneous attachment of myosin to actin; acetylcholine stimulates contraction skeletal muscle.

99) Name and describe the functions of the two contractile, two regulatory, and five structural proteins seen in skeletal muscle fibers.

Answer:
1. Contractile – myosin in thick filaments is motor protein; actin in thin filaments binds to myosin.
2. Regulatory proteins – troponin blocks myosin-binding sites on actin at rest; tropomyosin holds troponin in place.
3. Structural – titin stabilizes position of thick filaments; myomesin binds to titin and holds thick filaments together; nebulin helps maintain alignment of thin filaments; alpha-actinin forms latticework to hold actin, nebulin, and titin in Z disc; dystrophin links thin filaments to integral membrane proteins of sarcolemma to help transmit tension generated by sarcomeres to tendons.

100) Old Farmer Brown complains that at age 88, "I can't get around like I used to." What changes in his muscle tissue might make this true? What would you recommend to slow the process?

Answer: With age general loss of muscle mass occurs, and there is replacement of muscle by fibrous connective tissue and adipose tissue possibly due to progressive inactivity. Also, maximal strength of contraction decreases and reflexes are slowed. There is an increase in the number of slow oxidative fibers. One could recommend that he keep moving and lifting around the farm. Possibly creatine supplements could help.

CHAPTER 11   The Muscular System

MULTIPLE CHOICE. Choose the one alternative that best completes the statement or answers the question.

1) In a muscle group, the muscle that relaxes during a particular action is the
    A) prime mover.
    B) antagonist.
    C) fixator.
    D) synergist.
    E) Both B and D are correct.

    Answer: B

2) The insertion of a skeletal muscle is the
    A) connection to the bone that remains stationary while the muscle contracts.
    B) connection to the bone that moves while the muscle contracts.
    C) point at which effort is applied in an anatomical lever system.
    D) point at which the tendon attaches to the muscle itself.
    E) Both B and C are correct.

    Answer: E

3) Which of the following would allow a greater range of motion around a joint?
    A) having the insertion point of the muscle to be very close to the joint
    B) having the insertion point of the muscle to be as far as possible from the joint
    C) having very short fascicles
    D) Both B and C are correct.
    E) Range of motion is not related to the point of muscle insertion or fascicle length

    Answer: A

4) The masseter
    A) sucks in the cheeks.
    B) protracts the mandible.
    C) elevates the mandible.
    D) depresses the mandible.
    E) Both B and D are correct.

    Answer: E

5) Thoracic volume is increased during normal breathing by the
   A) internal intercostals.
   B) external obliques.
   C) diaphragm.
   D) trapezius.
   E) Both B and C are correct.

   Answer: C

6) ALL of the following move the mandible EXCEPT the
   A) masseter.
   B) buccinator.
   C) temporalis.
   D) digastric.
   E) lateral pterygoid.

   Answer: B

7) The tongue is protracted by the
   A) masseter.
   B) buccinator.
   C) platysma.
   D) styloglossus.
   E) genioglossus.

   Answer: E

8) Which of the following elevates the hyoid bone and depresses the mandible?
   A) styloglossus
   B) genioglossus
   C) risorius
   D) digastric
   E) sternocleidomastoid

   Answer: D

9) The neck is divided into anterior and posterior triangles by the
   A) sternocleidomastoid.
   B) masseter.
   C) trapezius.
   D) splenius capitis.
   E) semispinalis capitis.

   Answer: A

10) The anterior triangle of the neck contains ALL of the following EXCEPT the
   A) parotid salivary gland.
   B) brachial plexus.
   C) vagus nerve.
   D) internal jugular vein.
   E) submandibular lymph nodes.
   Answer: B

11) Jim has been lifting weights and doing exercises, such as sit-ups, to get that rippled "washboard" abdomen look. What muscle provides those "rippled abs?"
   A) transverse abdominis
   B) quadratus lumborum
   C) rectus abdominis
   D) latissimus dorsi
   E) pectoralis major
   Answer: C

12) The bone(s) that is (are) the origin of the rectus abdominis is (are) the
   A) ilium.
   B) ischium.
   C) pubis.
   D) sternum.
   E) last pair of vertebrochondral ribs.
   Answer: C

13) The bone(s) that is (are) the origin of the internal oblique and transverse abdominis is (are) the
   A) ilium.
   B) ischium.
   C) pubis.
   D) sternum.
   E) last pair of vertebrochondral ribs.
   Answer: A

14) The phrenic nerve stimulates the
   A) sternocleidomastoid.
   B) rectus abdominis.
   C) splenius capitis.
   D) diaphragm.
   E) external intercostals.
   Answer: D

15) During inspiration (i.e., breathing in), which of the following happens to increase the size of the thoracic cavity?

A) The abdominal muscles relax.

B) The internal intercostals contract to pull on the ribs.

C) The external intercostals relax to "release" the ribs.

D) The diaphragm contracts to flatten out.

E) All of the above are correct.

Answer: D

16) Which of the following is part of the pelvic diaphragm?

A) iliococcygeus

B) external urethral sphincter

C) bulbospongiosus

D) ischiocavernosus

E) Both A and C are correct.

Answer: A

17) The perineum is

A) the anterior of the two triangles separated by the sternocleidomastoid.

B) the floor of the thoracic cavity.

C) a band of fibrous tissue extending from the pubis to the sternum.

D) a group of muscles that move the foot.

E) the region of the trunk inferior to the pelvic diaphragm.

Answer: E

18) ALL of the following are muscles of the perineum EXCEPT the

A) iliococcygeus.

B) external urethral sphincter.

C) bulbospongiosus.

D) ischiocavernosus.

E) Both A and C are correct.

Answer: A

19) The muscle that helps maintain the erection of the penis and clitoris is the

A) external urethral sphincter.

B) deep transverse perineus.

C) ischiocavernosus.

D) coccygeus.

E) pubococcygeus.

Answer: C

20) ALL of the following move the scapula EXCEPT the
    A) rhomboideus major.
    B) serratus anterior.
    C) pectoralis major.
    D) trapezius.
    E) pectoralis minor.
    Answer: C

21) A hole in the linea alba that allows passage of the spermatic cord in males is the
    A) rotator cuff.
    B) inguinal canal.
    C) extensor retinaculum.
    D) levator ani.
    E) carpal tunnel.
    Answer: B

22) ALL of the following are part of the rotator cuff EXCEPT the
    A) teres minor.
    B) infraspinatus.
    C) supraspinatus.
    D) subscapularis.
    E) pectoralis major.
    Answer: E

23) The group of muscles known as the hamstrings includes the
    A) rectus femoris, vastus lateralis, and vastus medialis.
    B) sartorius and gracilis.
    C) gluteus maximus, medius, and minimus.
    D) external oblique, internal oblique, and rectus abdominis.
    E) biceps femoris, semitendinosus, and semimembranosus.
    Answer: E

24) The deltoid performs which of the following actions?
    A) adduction of the arm
    B) abduction of the arm
    C) flexion of the forearm
    D) extension of the forearm
    E) Both B and C are correct.
    Answer: B

25) The radial tuberosity is the insertion point for the
   A) triceps brachii.
   B) biceps brachii.
   C) brachialis.
   D) deltoid.
   E) Both A and B are correct.

   Answer: B

26) Which of the following is part of the flexor compartment of the arm?
   A) anconeus
   B) triceps brachii
   C) brachialis
   D) biceps brachii
   E) Both C and D are correct.

   Answer: E

27) The olecranon of the ulna is the insertion point for the
   A) triceps brachii.
   B) biceps brachii.
   C) brachialis.
   D) deltoid.
   E) Both A and B are correct.

   Answer: A

28) The deep fascia covering the palmar surface of the carpal bones is the
   A) linea alba.
   B) fascia lata.
   C) galea aponeurotica.
   D) flexor retinaculum.
   E) extensor retinaculum.

   Answer: D

29) The origin of a muscle refers to the
   A) embryonic derivation from a particular germ layer.
   B) attachment to the moving bone.
   C) attachment to the stationary bone.
   D) point at which the tendon meets the bone.
   E) point at which blood vessels and nerves enter the muscle.

   Answer: C

30) Which of the following refers to the relative size of a muscle?
   A) pectinate
   B) biceps
   C) vastus
   D) rectus
   E) quadratus
   Answer: C

31) Which of the following terms seen in muscle names means "shortest?"
   A) minimus
   B) brevis
   C) minor
   D) piriformis
   E) vastus
   Answer: B

32) A word in a muscle name that indicates that the muscle decreases the size of an opening is
   A) sphincter.
   B) extensor.
   C) levator.
   D) tensor.
   E) rotator.
   Answer: A

33) In a third class lever system, the arrangement of the system's components is such that the
   A) fulcrum is between the effort and the resistance.
   B) effort is between the fulcrum and the resistance.
   C) resistance is between the fulcrum and the effort.
   D) effort and the resistance are equidistant from the fulcrum.
   E) There is no fulcrum.
   Answer: B

34) The extrinsic muscles that move the wrist, hand, and digits originate on the
   A) humerus.
   B) proximal ends of the radius and ulna.
   C) carpals.
   D) distal ends of the radius and ulna.
   E) metacarpals.
   Answer: A

35) The hypothenar muscles act on the
    A) thumb.
    B) little finger.
    C) big toe.
    D) carpals.
    E) metatarsals.

    Answer: B

36) The nerve that passes through the carpal tunnel is the
    A) sciatic.
    B) phrenic.
    C) median.
    D) vagus.
    E) trigeminal.

    Answer: C

37) Which of the following originates on the transverse processes of the thoracic vertebrae?
    A) iliocostalis thoracis
    B) longissimus thoracis
    C) interspinales
    D) middle scalene
    E) All of these are correct.

    Answer: B

38) The head is laterally flexed and rotated by the
    A) iliocostalis thoracis
    B) longissimus thoracis
    C) interspinales
    D) middle scalene
    E) All of these are correct.

    Answer: D

39) The superior and inferior rectus muscles are prime movers for
    A) flexion and extension of the lower leg.
    B) flexion and extension of the thigh.
    C) elevation and depression of the eyelid.
    D) flexion and extension of the vertebral column.
    E) movements of the eyeball.

    Answer: E

40) Muscles with the combining form -glossus in the name cause movements of the
   A) eyeball.
   B) eyelid.
   C) tongue.
   D) vocal cords.
   E) ossicles in the ear.

   Answer: C

41) Which of the following muscles is deepest?
   A) rectus abdominis
   B) external oblique
   C) internal oblique
   D) transverse abdominis
   E) pectoralis major

   Answer: D

42) Which of the following muscles that move the humerus is designated as an axial muscle?
   A) latissmus dorsi
   B) deltoid
   C) teres minor
   D) subscapularis
   E) supraspinatus

   Answer: A

43) Which of the following would act as an antagonist to the tibialis anterior?
   A) rectus femoris
   B) biceps femoris
   C) tensor fasciae latae
   D) peroneus longus
   E) gracilis

   Answer: D

44) Which of the following acts as an antagonist to the rectus femoris?
   A) vastus lateralis
   B) semimembranosus
   C) gastrocnemius
   D) tibialis anterior
   E) adductor magnus

   Answer: B

45) Which of the following acts as an antagonist to the deltoid?
   A) biceps brachii
   B) triceps brachii
   C) pectoralis major
   D) sternocleidomastoid
   E) external intercostals

   Answer: C

46) Which of the following acts as an antagonist to the platysma?
   A) occipitalis
   B) trapezius
   C) sternocleidomastoid
   D) masseter
   E) orbicularis oculi

   Answer: D

47) In an anatomical lever system, the effort is the
   A) bone on which the muscle originates.
   B) joint.
   C) muscular contraction pulling on the insertion point.
   D) weight of the body part to be moved.
   E) brain activity that stimulates movement.

   Answer: C

48) Two structurally identical muscles cross a joint. Muscle X inserts one inch from the joint. Muscle Y inserts three inches from the joint. Which of the following statements is most likely to be TRUE?
   A) Muscle X will produce a stronger movement due to greater leverage.
   B) Muscle Y will produce a stronger movement due to greater leverage.
   C) Muscle Y will produce a stronger movement because the tendon must be longer.
   D) Muscle Y will produce a greater range of motion because the lever is longer.
   E) Muscles X and Y must produce movements of equal strength, regardless of point of insertion, because they are structurally identical.

   Answer: B

49) The most powerful abductor of the femur at the hip joint is the
   A) semitendinosus.
   B) quadriceps femoris.
   C) quadratus femoris.
   D) gluteus medius.
   E) gluteus maximus.

   Answer: D

50) The adductors of the thigh have their origins and insertions on the

   A) ilium and greater trochanter of the femur.

   B) ischial tuberosities and condyles of the femur.

   C) pubis and linea aspera of the femur.

   D) anterior inferior iliac spines and tibial tuberosities.

   E) pubis and tibial tuberosities.

   Answer: C

MATCHING. Choose the item in column 2 that best matches each item in column 1.

51)	Largest muscle.	A.	Vastus
52)	Shortest muscle.	B.	Latissimus
53)	Widest muscle.	C.	Magnus
54)	Large muscle.	D.	Maximus
55)	Hugh muscle.	E.	Brevis

   Answers: 51) D. 52) E. 53) B. 54) C. 55) A.

56)	Triangular muscle.	A.	Gracilis
57)	Saw-toothed muscle.	B.	Piriformis
58)	Diamond-shaped muscle.	C.	Serratus
59)	Pear-shaped muscle.	D.	Rhomboid
60)	Slender muscle.	E.	Deltoid

   Answers: 56) E. 57) C. 58) D. 59) B. 60) A.

61)	Increases the angle of a bone.	A.	Tensor
62)	Turns palm superiorly.	B.	Supinator
63)	Moves a bone closer to the midline.	C.	Levator
64)	Makes a body part rigid.	D.	Adductor
65)	Raises or elevates a body part.	E.	Extensor

   Answers: 61) E. 62) B. 63) D. 64) A. 65) C.

66)	Four origins to a muscle.	A.	Biceps
67)	Two origins to a muscle.	B.	Quadriceps
68)	Parallel to the midline.	C.	Rectus
69)	Diagonal to the midline.	D.	Transverse
70)	Perpendicular to the midline.	E.	Oblique

   Answers: 66) B. 67) A. 68) C. 69) E. 70) D.

SHORT ANSWER. Write the word or phrase that best completes the statement or answers the question.

71) The attachment of a muscle to the movable bone is called the _____.

Answer: insertion

72) In a third class lever, the _____ is between the _____ and the _____.

Answer: effort; resistance; fulcrum

73) When the resistance is close to the fulcrum and the effort is applied farther away, the lever operates at a mechanical _____.

Answer: advantage

74) The word used in muscle names that means "diagonal to the midline" is _____.

Answer: oblique

75) In anatomical lever systems, bones act as _____.

Answer: levers

76) In an anatomical lever system, the fulcrum is the _____.

Answer: joint

77) A muscle that causes a particular action is called the _____.

Answer: prime mover

78) When a prime mover contracts, the _____ relaxes.

Answer: antagonist

79) The word in a muscle name that means "produces superior movements" is _____.

Answer: levator

80) Muscles that help stabilize movements and help prime movers work more efficiently are called _____ and _____.

Answer: synergists; fixators

81) The muscle that closes and protrudes the lips is the _____.

Answer: orbicularis oris

82) The upper eyelid is elevated by the _____.

Answer: levator palpebrae superioris

83) The masseter originates on the maxilla and zygomatic arch and inserts onto the _____ of the _____.

Answer: angle and ramus; mandible

84) When general anesthesia is administered during surgery, a total relaxation of the _____ muscle results, allowing the tongue to fall posteriorly.

Answer: genioglossus

85) The sternocleidomastoid has its origins on the sternum and the _____.

Answer: clavicle

86) A tough fibrous band that extends from the xiphoid process to the pubic symphysis is the _____.

Answer: linea alba

87) The inferior free border of the external oblique aponeurosis forms the _____.

Answer: inguinal ligament

88) The floor of the thoracic cavity is formed by the _____.

Answer: diaphragm

89) The muscles that move the humerus that originate on the axial skeleton are the latissimus dorsi and the _____.

Answer: pectoralis major

90) The pubococcygeus and iliococcygeus muscles make up the _____ muscle.

Answer: levator ani

91) The triceps brachii inserts onto the _____ of the _____.

Answer: olecranon process; ulna

92) The region of the trunk inferior to the pelvic diaphragm is a diamond-shaped area called the _____.

Answer: perineum

93) The hamstrings are the biceps femoris, the semimembranosus, and the _____.

Answer: semitendinosus

94) The antagonists to the hamstrings regarding movements of the lower leg is the _____.

Answer: quadriceps femoris

95) The muscle that helps expel urine during urination, that helps propel semen along the urethra, and that assists in the erection of the penis is the _____ muscle.

Answer: bulbospongiosus

ESSAY. Write your answer in the space provided or on a separate sheet of paper.

96) Identify the anatomical parts corresponding to the generic components of a lever system. Describe the arrangement of these parts in first, second, and third class lever systems.

Answer: The bone is the lever; the joint is the fulcrum; the muscle contraction pulling on its insertion point is the effort; the weight of the part to be moved is the resistance. A first class lever has the fulcrum between the effort and the resistance; a second class lever has the resistance in the middle; the third class lever has the effort in the middle. In all cases, the lever moves around the fulcrum.

97) Discuss the roles of prime movers, antagonists, synergists, and fixators in movement.

Answer: The prime mover contracts to cause a particular action. The antagonist causes the opposite action, and so, must relax while the prime mover contracts. Synergists prevent unwanted movements during an action, while fixators stabilize the origin of the prime mover. Both allow the prime mover to work more efficiently.

98) Chuck has a rotator cuff injury. What muscles and associated structures might be involved, and what sorts of activities might have led up to this injury. What movements might be inhibited by this injury?

Answer: The tendons of the subscapularis, supraspinatus, infraspinatus, and teres minor make up the rotator cuff. Any activity involving these muscles could be the problem—from throwing baseballs to shoveling snow. Inhibited movements depend on specific muscle involved—medial and lateral rotation, adduction, abduction, or extension of arm.

99) Name the three groups of muscles constituting the intrinsic muscles of the hand. Briefly describe their structures and functions.

Answer:
1. Thenar—four muscles that act on the thumb and form the lateral rounded contour of the palm.
2. Hypothenar—three muscles that act on the little finger and form the medial rounded contour of the palm.
3. Intermediate—12 muscles acting on all digits except the thumb; subgrouped as lumbricals, palmar interossei, and dorsal interossei, and located between metacarpals; needed for all phalangeal movements.

100) Name and describe the locations and actions of the muscles typically used in breathing.

Answer: The diaphragm forms the floor of the thoracic cavity. It flattens during inspiration to increase the size of the thoracic cavity. The external intercostals between the ribs increase the lateral and anteroposterior dimensions of the thorax. Internal intercostals between ribs pull ribs together in the opposite movement during expiration to decrease the size of the thoracic cavity. The diaphragm relaxes during expiration to form a dome and decrease the size of the thoracic cavity.

CHAPTER 12   Nervous Tissue

MULTIPLE CHOICE. Choose the one alternative that best completes the statement or answers the question.

1) Small masses of neuron cell bodies located outside the CNS are called
   A) interneurons.
   B) plexuses.
   C) nerves.
   D) ganglia.
   E) nuclei.

   Answer: D

2) Collections of nerve cell bodies outside the central nervous system are called
   A) nuclei.
   B) nerves.
   C) ganglia.
   D) tracks.

   Answers: C

3) The substance released at axonal endings to propagate a nervous impulse is called a
   A) ion.
   B) cholinesterase.
   C) neurotransmitter.
   D) biogenic amine.

   Answer: C

4) Neuroglia in the central nervous system that produce the myelin sheath are the
   A) astrocytes.
   B) oligodendrocytes.
   C) microglia.
   D) ependymal cells.
   E) Schwann cells.

   Answer: B

5) Neuroglia that are positioned between neurons and capillaries to form part of the blood-brain barrier are the

A) astrocytes.

B) oligodendrocytes.

C) microglia.

D) ependymal cells.

E) Schwann cells.

Answer: A

6) The autonomic nervous system is divided into

A) somatic and sympathetic nervous systems.

B) afferent and efferent nervous systems.

C) sympathetic and parasympathetic nervous systems.

D) central nervous system and peripheral nervous system.

Answer: C

7) An axolemma is a

A) plasma membrane.

B) nerve cytoplasm.

C) endoplasmic reticulum.

D) Golgi apparatus.

Answer: A

8) Nissl bodies are

A) the microtubules involved in fast axonal transport.

B) clusters of rough ER in the cell body.

C) gaps in the myelin sheath.

D) clusters of cell bodies in the CNS.

E) clusters of cell bodies in the PNS.

Answer: B

9) Synaptic vesicles store

A) glycogen for energy production.

B) lipofuscin.

C) neurotransmitter.

D) calcium ions.

E) enzymes fore degrading neurotransmitter.

Answer: C

10) The site of communication between two neurons is called a
   A) Nissl body.
   B) synapse.
   C) varicosity.
   D) dendrite.
   E) All are correct.

   Answer: B

11) Neurons that have one main dendrite and one axon are called
   A) multipolar neurons.
   B) bipolar neurons.
   C) unipolar neurons.
   D) monopolar neurons.
   E) neuroglia.

   Answer: B

12) Gaps in the myelin sheath are called
   A) oligodendrocytes.
   B) Nodes of Ranvier.
   C) Sheath of Schwann.
   D) gliomas.

   Answer: B

13) The function of Schwann cells is to
   A) form the myelin sheaths of neurons in the PNS.
   B) form the myelin sheaths of neurons in the CNS.
   C) act as part of the blood-brain barrier.
   D) act as interneurons.
   E) produce cerebrospinal fluid.

   Answer: A

14) White matter of the nervous system is composed of
   A) aggregations of myelinated axons.
   B) aggregations of cell bodies.
   C) aggregations of nuclei of cell bodies.
   D) aggregation of ganglia.
   E) All are correct.

   Answer: A

15) The Nodes of Ranvier are
   A) sites of neurotransmitter storage.
   B) where the nucleus of a neuron is located.
   C) gaps in the myelin sheath.
   D) sites of myelin production.
   E) sites of neurotransmitter production.

   Answer: C

16) The branch of a neuron that carries a nerve impulse away from the cell body is the
   A) axon.
   B) dendrite.
   C) perikaryon.
   D) Nissl body.
   E) Node of Ranvier.

   Answer: A

17) Ion channels open and close due to the presence of
   A) ligands.
   B) lids.
   C) gates.
   D) doors.

   Answer: C

18) Afferent nerves conduct nerve impulses from
   A) the central nervous system to effectors.
   B) effectors to the central nervous system.
   C) receptors to the central nervous system.
   D) the central nervous system to receptors.
   E) one effector to another.

   Answer: C

19) The absolute refractory period is the time
   A) it takes a neurotransmitter to diffuse across a synaptic cleft.
   B) it takes to reach threshold via temporal summation.
   C) following birth when neurons can still reproduce.
   D) between injury to an axon and recovery of function.
   E) following an action potential during which a second action potential cannot be initiated regardless of stimulus strength.

   Answer: E

20) During the depolarization phase of an action potential, which of the following is the primary activity?

A) Potassium ions are flowing into the cell.

B) Potassium ions are flowing out of the cell.

C) Sodium ions are flowing into the cell.

D) Sodium ions are flowing out of the cell.

E) Neurotransmitter is diffusing into the cell.

Answer: C

21) During the depolarization phase of an action potential, which of the following situations exists?

A) The inside of the membrane is becoming more negative with respect to the outside.

B) The inside of the membrane is becoming more positive with respect to the outside.

C) The membrane is becoming less permeable to all ions.

D) The membrane potential remains constant.

E) The threshold is changing.

Answer: B

22) During the repolarization phase of an action potential, which of the following is the primary activity?

A) Potassium ions are flowing into the cell.

B) Potassium ions are flowing out of the cell.

C) Sodium ions are flowing into the cell.

D) Sodium ions are flowing out of the cell.

E) The membrane is impermeable to all ions.

Answer: B

23) A resting membrane potential is due to

A) extracellular fluid rich in sodium.

B) intracellular fluid rich in potassium.

C) plasma membrane that is more permeable to potassium than sodium.

D) All are correct.

Answer: D

24) When ion movements across the membrane occur during the depolarization and repolarization phases of an action potential, they are moving by

A) primary active transport.

B) secondary active transport.

C) exocytosis.

D) filtration.

E) simple diffusion.

Answer: E

25) The after-hyperpolarization that follows an action potential occurs due to a rapid
   A) outflow of sodium ions.
   B) outflow of potassium ions.
   C) influx of sodium ions.
   D) influx of potassium ions.
   E) release of neurotransmitter.

   Answer: B

26) The factor most affecting the rate of impulse conduction is the
   A) length of the axon.
   B) site of initial stimulation.
   C) rate of axonal transport.
   D) number of neuroglia associated with it.
   E) presence or absence of a myelin sheath.

   Answer: E

27) The term *saltatory conduction* refers to
   A) "leaping" of an action potential across a synapse.
   B) movement of sodium ions into the cell during depolarization.
   C) one-way conduction of a nerve impulse across a synapse.
   D) conduction of a nerve impulse along a myelinated axon.
   E) action of the sodium-potassium pump.

   Answer: D

28) The period of time during which a second action potential can be initiated is called
   A) absolute refractory period.
   B) all of nothing.
   C) relative refractory period.
   D) excitatory refractory period.

   Answer: C

29) The effect of a neurotransmitter on the postsynaptic cell occurs when the neurotransmitter
   A) diffuses into the postsynaptic cell.
   B) flows along the postsynaptic cell membrane into transverse tubules.
   C) is broken down by enzymes in the synaptic cleft.
   D) actively transports sodium ions into the postsynaptic cell.
   E) binds to specific receptors on the postsynaptic cell membrane.

   Answer: E

30) A local anesthetic functions by blocking
   A) the opening of potassium channels.
   B) the opening of sodium channels.
   C) the closing of potassium channels.
   D) the closing of sodium channels.

   Answer: B

31) The type A neurons
   A) are all myelinated.
   B) are the largest neurons in diameter.
   C) exhibit saltatory conduction.
   D) carry motor neurons.
   E) All answers are correct.

   Answer: E

32) What would happen to a postsynaptic neuron that binds a neurotransmitter that closes potassium ion channels?
   A) An EPSP would result because potassium ions are positively charged, and as they accumulate inside the cell, the membrane moves closer to threshold.
   B) An EPSP would result because the sodium/potassium pump would work to force potassium ions out of the cell, and sodium ions would be brought in as a result.
   C) An IPSP would result because potassium ions are negatively charged, and as they accumulate inside the cell, the membrane becomes more hyperpolarized.
   D) There would be no change in membrane potential because the membrane potential is dependent only on the opening and closing of sodium ion channels.
   E) There would be no change in membrane potential because there would be an increase in production of intracellular anions to compensate for the added potassium ions.

   Answer: A

33) A neurotransmitter that allows sodium ions to leak into a postsynaptic neuron causes
   A) excitatory postsynaptic potentials.
   B) inhibitory postsynaptic potentials.
   C) no changes in resting potential.
   D) an alteration of the membrane threshold.
   E) damage to the myelin sheath.

   Answer: A

34) The depolarizing phase of nerve impulse opens voltage gated calcium channels in the
    A) membrane of synaptic end bulbs.
    B) membrane of post-synaptic membranes.
    C) synaptic cleft.
    D) All answers are correct.

    Answer: A

35) When depolarization reaches the axon terminal of a presynaptic neuron, the next event is
    A) immediate release of neurotransmitter.
    B) uptake of neurotransmitter from the synaptic cleft.
    C) diffusion of calcium ions out of the cell.
    D) diffusion of calcium ions into the cell.
    E) active transport of calcium ions out of the cell.

    Answer: D

36) Most excitatory neurons in the CNS communicate via the neurotransmitter
    A) acetylcholine.
    B) epinephrine.
    C) GABA.
    D) glutamate.
    E) dopamine.

    Answer: D

37) Summation results from buildup of neurotransmitter released simultaneously by several presynaptic neurons, this is called
    A) spatial summation.
    B) temporal summation.
    C) federal summation.
    D) enzymatic summation.

    Answer: A

38) A drug that is an MAO inhibitor would
    A) block the action of acetylcholine.
    B) increase the effects of catecholamines by inhibiting their breakdown.
    C) reduce the effects of catecholamines by promoting their breakdown.
    D) block the manufacture of acetylcholine.
    E) block the manufacture of epinephrine.

    Answer: B

39) The neurotransmitter that is not synthesized in advance and packaged into synaptic vesicles is
   A) acetylcholine.
   B) GABA.
   C) nitric oxide.
   D) dopamine.
   E) epinephrine.

   Answer: C

40) The plasma membrane of a neuron is more permeable to potassium ions than to sodium ions because the membrane has
   A) more voltage-gated sodium ion channels.
   B) more ligand-gated potassium ion channels.
   C) more potassium leakage channels.
   D) fewer voltage-gated sodium ion channels.
   E) more carrier molecules for potassium ions.

   Answer: C

41) Nitric oxide is thought to work by
   A) opening voltage-gated sodium ion channels.
   B) opening ligand-gated potassium ion channels.
   C) opening ligand-gated chloride ion channels.
   D) preventing the breakdown of catecholamines.
   E) activating an enzyme for production of cyclic GMP.

   Answer: E

42) Catecholamines include the following:
   A) adrenalin.
   B) norepinephrine.
   C) dopamine.
   D) All are correct answers.

   Answer: D

43) The threshold of a neuron is the
   A) time between binding of the neurotransmitter and firing of an action potential.
   B) voltage at which the inflow of sodium ions causes reversal of the resting potential.
   C) total number of sodium ions that enters the cell before sodium inactivation gates close.
   D) total amount of neurotransmitter it takes to cause an action potential.
   E) voltage across the resting cell membrane.

   Answer: B

44) Parkinson's disease occurs when there has been degeneration of the axons containing
   A) acetylcholine.
   B) GABA.
   C) endorphins.
   D) dopamine.
   E) epinephrine.

   Answer: D

45) Endorphins and enkephalins are neuropeptides that
   A) induce pain.
   B) relieve pain.
   C) block the action of serotonin.
   D) are not stored in synaptic vesicles.
   E) regulate water balance.

   Answer: B

46) 5-hydroxytryptamine (5HT) is also called
   A) epinephrine.
   B) dopamine.
   C) serotonin.
   D) monoamine oxidase.

   Answer: C

47) A typical pain neurotransmitter is
   A) substance P.
   B) nitric oxide.
   C) enkephalin.
   D) serotonin.

   Answer: A

48) Wallerian degeneration is degeneration of
   A) neuroglia in the blood-brain barrier.
   B) Schwann cells.
   C) the myelin sheath in multiple sclerosis.
   D) unnecessary neurons during fetal development.
   E) the distal fragment of a severed neuronal process and its myelin sheath.

   Answer: E

49) Following injury to a peripheral neuron, chromatolysis occurs, which is
    A) release of neurotransmitter from remaining synaptic vesicles.
    B) breakup of the Nissl bodies.
    C) degeneration of the myelin sheath.
    D) reproduction of the local neuroglia.
    E) degeneration of the distal fragment of a severed neuronal process and its myelin sheath.

    Answer: B

50) What part of the neuron is necessary for regeneration to occur after injury?
    A) neurolemma
    B) axon
    C) dendrite
    D) myelin sheath

    Answer: A

MATCHING. Choose the item in column 2 that best matches each item in column 1.

51) Nucleus containing.	A. Endoplasmic reticulum
52) Nissl body.	B. Axoplasma
53) Cytoplasm.	C. Neurolemma
54) Plasma membrane.	D. Cytoskeleton
	E. Cell body

Answers: 51) E. 52) A. 53) B. 54) C.

55) The cell body of a neuron.	A. Gray matter
56) Myelinated axons.	B. Neuroglia
57) The "glue" that holds the nervous tissue.	C. White matter
58) A group of cell bodies.	D. Ganglia
	E. Glioma

Answers: 55) A. 56) C. 57) B. 58) D.

59) During a resting membrane potential.	A. Potassium leaks out
60) During an action potential.	B. Sodium leaks out
61) Inhibitory postsynaptic potential.	C. Sodium gates open first
62) Presynaptic membrane.	D. Potassium gates open first
	E. Calcium gates open

Answers: 59) A. 60) C. 61) D. 62) E.

63)	Neurotransmitter at neuromyojunctions.	A.	Norepinephrine
64)	Important inhibitory neurotransmitters.	B.	Acetylcholine
65)	Associated with Parkinson's disease.	C.	Gamma aminobutyric acid
66)	5-hydroxytryptamine (5-HT).	D.	Dopamine
		E.	Serotonin

Answer: 63) B. 64) C. 65) D. 66) E.

67)	Progressive destruction of myelin sheaths.	A.	Neuropathy
68)	Disorder of a cranial nerve.	B.	Rabies
69)	Caused by a virus in the brain.	C.	Multiple sclerosis
70)	Multiple abnormal action potentials.	D.	Guillain-Barre Syndrome
		E.	Epilepsy

Answer: 67) C. 68) A. 69) B. 70) E.

SHORT ANSWER. Write the word or phrase that best completes each statement or answers the question.

71) The two principal divisions of the nervous system are the _____ and the _____.
Answer: central nervous system; peripheral nervous system

72) Neurons that serve the integrative function of the nervous system are called _____.
Answer: interneurons (association neurons)

73) The _____ nervous system provides involuntary regulation of the gastrointestinal tract.
Answer: enteric

74) _____ neurons transmit nerve impulses from receptors to the central nervous system.
Answer: Afferent

75) Motor neurons that conduct impulses from the CNS to smooth and cardiac muscle belong to the _____ nervous system.
Answer: autonomic

76) Neuroglia that produce the myelin sheath in the central nervous system are the _____.
Answer: oligodendrocytes

77) The nucleus of a neuron is located in the _____ of a neuron.
Answer: cell body

78) Nerve impulses are conducted toward the cell body by a neuronal process called a(n) _____.
Answer: dendrite

79) Neurons having one main axon and one main dendrite are classified structurally as being _____.

Answer: bipolar

80) The outer nucleated cytoplasmic layer of a Schwann cell is the _____.

Answer: neurolemma

81) White matter looks white in color due to the presence of _____.

Answer: myelin

82) A cluster of neuron cell bodies within the CNS is called a(n) _____.

Answer: nucleus

83) The effectors for general somatic efferent neurons are _____.

Answer: skeletal muscles

84) Neuroglia of the neurolemma are the _____.

Answer: oligodendrocytes

85) Ion channels in a plasma membrane that are always open are called _____ channels.

Answer: leakage

86) A membrane whose polarization is more negative than the resting level is said to be _____.

Answer: hyperpolarized

87) Ion channels in a plasma membrane that open in response to changes in membrane potential are called _____ channels.

Answer: voltage-gated

88) Recovery of the resting potential due to opening of voltage-gated potassium ion channels and closing of voltage-gated sodium ion channels is called _____.

Answer: repolarization

89) Impulse conduction that appears to jump from one Node of Ranvier to the next is called _____ conduction.

Answer: saltatory

90) Ion channels in a plasma membrane that open or close in response to a specific chemical stimulus are called _____ channels.

Answer: ligand-gated

91) Ion channels in a plasma membrane that open in response to vibration or pressure are called _____ channels.

Answer: mechanically gated

92) At rest, the extracellular fluid around a neuron's plasma membrane is especially rich in positively charged _____ ions.

Answer: sodium

93) In the cytosol, the predominant cation is _____.

Answer: potassium

94) A neurotransmitter that causes hyperpolarization of the membrane is said to cause a(n) _____ postsynaptic potential.

Answer: inhibitory

95) The critical membrane potential at which a neuron must generate an action potential is called _____.

Answer: threshold

ESSAY. Write your answer in the space provided or on a separate sheet of paper.

96) Describe the characteristics of the neuron cell membrane and its environment that contribute to the existence of a resting potential.

Answer: Sodium and chloride ions predominate in the ECF, while potassium, organic phosphates, and amino acids predominate in the ICF. The membrane is moderately permeable to potassium but only slightly permeable to sodium. Any sodium that leaks in is removed via active transport pumps, which pump out three sodium ions for each two potassium ions imported. Anions in the ICF are generally too large to escape. The net effect is an accumulation of positive charges outside, while the inner surface of the membrane becomes more negatively charged.

97) Describe the phases of an action potential. including all appropriate ion movements and the mechanisms by which such movements occur.

Answer:
1. Depolarization—graded potential brings the membrane to threshold and voltage-gated sodium ion channels open; sodium rushes in by diffusion and creates positive feedback situation; sodium inactivation gates close just after activation gates open.
2. Repolarization—voltage-gated potassium ion channels open as soon as sodium ions channels are closing; potassium diffuses out to restore resting potential; voltage-gated sodium ion channels revert to resting state.
3. After-hyperpolarization—may occur as large outflow of potassium passes normal resting potential.

98) Describe the means by which a neurotransmitter may be removed from the synaptic cleft.

Answer:
1. simple diffusion away from cleft.
2. inactivation by enzymes.
3. uptake into presynaptic neuron or into neighboring neuroglia.

99) Bacteria produce a toxin that causes a flaccid paralysis. How might this toxin cause its effects?

Answer: It could block ACh release, increase acetylcholinesterase production, or block receptors so ACh cannot bind.

100) Bacteria produce a toxin that causes all skeletal muscles to contract at the same time. How might this toxin cause its effects?

Answer: It could block the action of acetylcholinesterase, mimic the action of ACh on postsynaptic receptors, or block glycine release (allows antagonists to contract with prime movers).

CHAPTER 13   The Spinal Cord and Spinal Nerves

MULTIPLE CHOICE. Choose the one alternative that best completes the statement or answers the question.

1) What would normally be found within the central canal of the spinal cord?
   A) blood
   B) myelin
   C) cerebrospinal fluid
   D) air
   E) gray matter

   Answer: C

2) The filum terminale is
   A) the roots of spinal nerves hanging inferiorly from the inferior end of the spinal cord in the vertebral column.
   B) an indentation on the dorsal side of the spinal cord.
   C) the tapered end of the spinal cord.
   D) an extension of the pia mater that anchors the spinal cord to the coccyx.
   E) where the cell bodies of sensory neurons are located.

   Answer: D

3) The cauda equina is
   A) the roots of spinal nerves hanging inferiorly from the inferior end of the spinal cord in the vertebral column.
   B) an indentation on the dorsal side of the spinal cord.
   C) the tapered end of the spinal cord.
   D) an extension of the pia mater that anchors the spinal cord to the coccyx.
   E) where the cell bodies of sensory neurons are located.

   Answer: A

4) The dorsal root ganglion is
   A) the roots of spinal nerves hanging inferiorly from the inferior end of the spinal cord in the vertebral column.
   B) an indentation on the dorsal side of the spinal cord.
   C) the tapered end of the spinal cord.
   D) an extension of the pia mater that anchors the spinal cord to the coccyx.
   E) where the cell bodies of sensory neurons are located.

   Answer: E

5) What would normally be found immediately surrounding central canal of the spinal cord?
   A) white matter
   B) gray matter
   C) cerebrospinal fluid
   D) the pia mater
   E) the dura mater
   Answer: B

6) Cerebrospinal fluid normally circulates in the
   A) epidural space.
   B) subdural space.
   C) subarachnoid space.
   D) ascending tracts.
   E) descending tracts.
   Answer: C

7) In the adult, the spinal cord extends from the medulla to the
   A) coccyx.
   B) sacral promontory.
   C) point of attachment of the most inferior pair of ribs.
   D) sacral hiatus.
   E) upper border of vertebra L2.
   Answer: E

8) Spinal nerves are considered mixed, which means that
   A) they contain both nerves and tracts.
   B) they contain both gray and white matter.
   C) they contain both afferent and efferent nerves.
   D) they use multiple types of neurotransmitters.
   E) a single nerve arises from multiple segments of the spinal cord.
   Answer: C

9) The part of a spinal nerve that contains only efferent fibers is the
   A) dorsal root.
   B) ventral root.
   C) dorsal ramus.
   D) ventral ramus.
   E) plexus.
   Answer: B

10) To do a lumbar puncture, the needle is inserted into the
   A) central canal.
   B) sacral plexus.
   C) nucleus pulposus.
   D) subarachnoid space.
   E) gray commissure.

   Answer: D

11) The nerve that stimulates the diaphragm to contract arises from the
   A) cervical plexus.
   B) lumbar plexus.
   C) brachial plexus.
   D) sacral plexus.
   E) intercostal nerves.

   Answer: A

12) The nerve that stimulates the diaphragm to contract is the
   A) median nerve.
   B) phrenic nerve.
   C) sciatic nerve.
   D) radial nerve.
   E) second intercostal nerve.

   Answer: B

13) The sacral plexus is the origin of the
   A) axillary nerve.
   B) obturator nerve.
   C) femoral nerve.
   D) sciatic nerve.
   E) Both C and D are correct.

   Answer: D

14) ALL of the following arise from the brachial plexus EXCEPT the
   A) axillary nerve.
   B) phrenic nerve.
   C) radial nerve.
   D) long thoracic nerve.
   E) median nerve.

   Answer: B

15) The endoneurium is the
    A) lining of the central canal of the spinal cord.
    B) space between the pia mater and the spinal cord.
    C) connective tissue surrounding an individual axon.
    D) connective tissue surrounding an entire nerve.
    E) group of neurons forming a spinal tract.

Answer: C

16) Spinal nerves emerge from the vertebral column via
    A) vertebral bodies.
    B) intervertebral discs.
    C) intervertebral foramina.
    D) vertebral foramina.
    E) the central canal.

Answer: C

17) If the ventral root of a spinal nerve was cut; there would be a complete loss of
    A) sensation.
    B) movement.
    C) sensation and movement.
    D) sensation, movement and autonomic activity.

Answer: B

18) The spinal cord passes through the
    A) vertebral bodies.
    B) intervertebral discs.
    C) intervertebral foramina.
    D) vertebral foramen.
    E) All of the above except the intervertebral foramina.

Answer: D

19) The spinal cord is suspended in the middle of its dural sheath by
    A) ascending spinal tracts.
    B) the cauda equina.
    C) cerebrospinal fluid.
    D) denticulate ligaments.
    E) epidural fat.

Answer: D

20) The diameter of the spinal cord is slightly larger in the
   A) cervical and thoracic regions.
   B) cervical and lumbar regions.
   C) cervical and sacral regions.
   D) thoracic and lumbar regions.
   E) lumbar and sacral regions.

   Answer: B

21) The subdural space normally contains
   A) fat.
   B) interstitial fluid.
   C) cerebrospinal fluid.
   D) blood.
   E) air.

   Answer: B

22) Neurons that transmsit impulses from the receptors to the central nervous system are called
   A) motor neurons.
   B) association neurons.
   C) bipolar neurons.
   D) sensory neurons.
   E) efferent neurons.

   Answer: D

23) The innermost layer of the meninges is the
   A) dura mater.
   B) arachnoid.
   C) pia mater.
   D) gray commissure.
   E) conus medullaris.

   Answer: C

24) The nerves to and from the lower limbs arise from the
   A) intercostal region.
   B) filum terminale.
   C) dura mater.
   D) brachial plexus.
   E) lumbar enlargement.

   Answer: E

25) A groove that penetrates the white matter of the spinal cord and divides it into right and left side is called

A) anterior median fissure.

B) posterior median fissure.

C) anterior median sulcus.

D) posterior median sulcus.

Answer: A

26) Ascending tracts contain

A) motor neurons.

B) sensory neurons.

C) cerebrospinal fluid.

D) only cell bodies.

E) only unmyelinated axons.

Answer: B

27) The spinothalamic tracts convey

A) motor impulses to skeletal muscles.

B) sensory impulses regarding proprioception.

C) sensory impulses regarding discriminative touch.

D) sensory impulses regarding vibration.

E) sensory impulses regarding temperature.

Answer: E

28) The cell bodies of the somatic motor neurons and motor nuclei are contained in the

A) anterior gray horns.

B) posterior gray horns.

C) anterior white commissures.

D) posterior white commissures.

E) ventral white commissures.

Answer: B

29) Which of the following lists the anatomical features in correct order from outermost to innermost?

A) dura mater, epidural space, arachnoid, subdural space, pia mater, subarachnoid space

B) pia mater, epidural space, dura mater, subdural space, arachnoid, subarachnoid space

C) arachnoid, subarachnoid space, pia mater, epidural space, dura mater, subdural space

D) epidural space, dura mater, subdural space, arachnoid, subarachnoid space, pia mater

E) dura mater, epidural space, subdural space, arachnoid, subarachnoid space, pia mater

Answer: D

30) Voluntary skeletal muscle movements are stimulated via the
   A) spinothalamic tracts.
   B) posterior column tracts.
   C) corticospinal tracts.
   D) vestibulospinal tract.
   E) tectospinal tract.

   Answer: C

31) The spinal cord is continuous with the
   A) occipital bone.
   B) cerebral cortex.
   C) medulla oblongata.
   D) thalamus.
   E) coccyx.

   Answer: C

32) ALL of the following would be examples of autonomic reflexes EXCEPT
   A) rising heart rate as blood pressure drops.
   B) contraction of smooth muscle in the wall of the gastrointestinal tract to push food along.
   C) secretion of hormones from the adrenal medulla during stress.
   D) contraction of the quadriceps femoris when the patellar tendon is stretched.
   E) dilation of blood vessels in the skin when body temperature increases.

   Answer: D

33) Cell bodies of motor neurons to skeletal muscles are located in the
   A) anterior gray horns.
   B) lateral gray horns.
   C) anterior white columns.
   D) posterior white columns.
   E) central canal.

   Answer: A

34) Sensory impulses are carried to the brain in the
   A) anterior spinal columns.
   B) posterior spinal columns.
   C) anterior spinal gray horns.
   D) posterior spinal gray horns.

   Answer: B

35) Which of the following lists the components of a reflex arc in the correct order of functioning?

A) receptor, motor neuron, integrating center, sensory neuron, effector

B) motor neuron, receptor, integrating center, sensory neuron, effector

C) receptor, sensory neuron, effector, motor neuron, integrating center

D) receptor, sensory neuron, integrating center, motor neuron, effector

E) effector, sensory neuron, integrating center, motor neuron, receptor

Answer: D

36) The following are ALL motor pathways EXCEPT

A) corticospinal.

B) corticobulbar.

C) spinothalamic.

D) rubrospinal.

E) vestibulospinal.

Answer: C

37) The contralateral reflex that helps you maintain your balance when the flexor reflex is initiated is the

A) stretch reflex.

B) tendon reflex.

C) crossed extensor reflex.

D) abdominal reflex.

E) patellar reflex.

Answer: C

38) Autonomic visceral reflexes involve

A) cardiac muscle.

B) small intestine smooth muscle.

C) glands of the stomach.

D) urinary bladder smooth muscle.

E) All answers are correct.

Answer: E

39) Slight stretching of the skeletal muscle spindles will initiate a(n)

A) autonomic reflex.

B) somatic reflex arc.

C) muscle contraction.

D) visceral reflex.

E) All answers are correct.

Answer: B

40) Stepping on a tack may stimulate the

    A) crossed extensor reflex.

    B) flexor reflex.

    C) spinal reflex.

    D) intersegmental reflexs.

    E) All answers are correct.

Answer: E

41) A spinal nerve is covered from the inside out with a

    A) myelin sheath, endoneurium, perineurium, epineurium.

    B) myelin sheath, epineurium, perineurium, endoneurium.

    C) myelin sheath, perineurium, epineurium, endoneurium.

    D) epineurium, perineurium, endoneurium.

Answer: A

42) In an adult, curling under of all the toes in response to gentle stroking of the lateral outer margin of the sole is

    A) normal.

    B) a positive Babinski sign.

    C) indicative of damage to the corticospinal tract.

    D) Both A and B are correct.

    E) Both B and C are correct.

Answer: A

43) The branch of a spinal nerve that supplies the vertebrae and blood vessels of the spinal cord is the

    A) anterior ramus.

    B) posterior ramus.

    C) meningeal branch.

    D) rami communicantes.

    E) endoneurium.

Answer: C

44) The branches of spinal nerves that form networks called plexuses are the

    A) dorsal roots.

    B) ventral roots.

    C) posterior rami.

    D) anterior rami.

    E) rami communicantes.

Answer: D

45) Inability to abduct and adduct the fingers and atrophy of the interosseous muscles of the hand indicate injury to the

A) axillary nerve.

B) phrenic nerve.

C) long thoracic nerve.

D) median nerve.

E) ulnar nerve.

Answer: E

46) Inability to adduct the leg indicates injury to the

A) femoral nerve.

B) pudendal nerve.

C) obturator nerve.

D) deep peroneal nerve.

E) tibial nerve.

Answer: C

47) The gastrocnemius and soleus are stimulated by the

A) obturator nerve.

B) median nerve.

C) superficial peroneal nerve.

D) deep peroneal nerve.

E) tibial nerve.

Answer: E

48) The plexus from which the sciatic nerve arises is located

A) between the psoas major and quadratus lumborum muscles.

B) anterior to the sacrum.

C) posterior to the sacrum.

D) in the obturator foramen.

E) between the gluteus maximus and gluteus medius muscles.

Answer: B

49) The one cranial nerve that has a dermatome is the

A) optic.

B) vagus.

C) trigeminal.

D) facial.

E) glossopharyngeal.

Answer: C

50) The polio virus typically attacks
  A) posterior root ganglia.
  B) neurons in the anterior gray horns of the spinal cord.
  C) only sensory neurons.
  D) only interneurons.
  E) the connective tissues surrounding neurons.

  Answer: B

MATCHING. Choose the item in column 2 that best matches each item in column 1.

51) Femoral nerve.
52) Median nerve.
53) Radial nerve.
54) Axillary nerve.

A. Triceps brachii
B. Deltoid
C. Extensors of the leg
D. Flexors of the forearm
E. Diaphragm

Answers: 51) C. 52) D. 53) A. 54) B.

55) Intercostal nerves.
56) Phrenic nerve.
57) Deep peroneal nerve.
58) Oburator nerve.

A. Adductors of the leg
B. Extensors of the leg
C. Diaphragm
D. Tibialis anterior
E. Abdominal muscles

Answers: 55) E. 56) C. 57) D. 58) A.

59) Extension of the pia mater attaching the spinal cord to the coccyx.
60) Contains cerebrospinal fluid.
61) Tapered end of the spinal cord below the lumbar enlargement.
62) Interstitital fluid.

A. Subarachnoid space
B. Conus medullaris
C. Filum terminale
D. Epidural space
E. Subdural space

Answers: 59) C. 60) A. 61) C. 62) A.

63) Middle layer of the meninges.
64) Outer layer of the brain.
65) Spinothalamic pathway.
66) Outer layer of the meninges.

A. Pia mater
B. Dura mater
C. Arachnoid
D. Ascending tracks
E. Descending tracks

Answers: 63) C. 64) A. 65) D. 66) B.

67) Attacks of pain along a sensory nerve.   A.   Meningitis
68) Inflammation of a nerve due to a bone fracture. B.   Paresthesia
69) Abnormal sensation such as tickling.   C.   Neuritis
70) Loss of sensation to a neuron.   D.   Neuralgia
                                                    E.   Nerve block

Answers: 67) D. 68) C. 69) B. 70) E.

SHORT ANSWER. Write the word or phrase that best completes each statement or answers the question.

71) The innermost of the meninges is the _____.

Answer: pia mater

72) Cerebrospinal fluid circulates in the _____ space.

Answer: subarachnoid

73) In the adult, the spinal cord extends from the medulla to the _____ vertebra.

Answer: second lumbar

74) The avascular layer of the meninges is the _____.

Answer: arachnoid

75) The extensions of the pia mater that suspend the spinal cord in the middle of the dural sheath are called _____.

Answer: denticulate ligaments

76) The cell bodies of peripheral sensory neurons are located in swellings known as _____.

Answer: posterior (dorsal) root ganglia

77) Motor neuron axons are contained in the _____ root of a spinal nerve.

Answer: ventral

78) Bundles of axons in the spinal cord having a common origin or destination and carrying similar information are called _____.

Answer: tracts

79) Sensory spinal tracts are also known as _____ tracts.

Answer: ascending

80) Proprioception and discriminative touch are some of the sensory inputs transmitted by the sensory tracts known as the _____ tract.

Answer: posterior column

81) The indirect descending pathways include the rubrospinal, tectospinal, and _____ tracts.

Answer: vestibulospinal

82) Changes in the internal or external environment are sensed by the component of a reflex arc known as the _____.

Answer: receptor

83) Reflexes involving smooth muscle, cardiac muscle, and glands are called _____ reflexes.

Answer: autonomic (visceral)

84) In a somatic reflex, the effector is a(n) _____.

Answer: skeletal muscle

85) A neural circuit which simultaneously causes contraction of one muscle and relaxation of its antagonists is called _____.

Answer: reciprocal innervation

86) The connective tissue covering a whole nerve is called the _____.

Answer: epineurium

87) The anterior rami of spinal nerves (except T2-T12) form networks called _____.

Answer: plexuses

88) The phrenic nerve stimulates the _____.

Answer: diaphragm

89) There are _____ pairs of spinal nerves.

Answer: 31

90) Wrist drop, the inability to extend the wrist and fingers, results from damage to the _____ nerve, which arises from the _____ plexus.

Answer: radial (or axillary); brachial

91) The branch of a spinal nerve containing the autonomic components is the _____.

Answer: rami communicantes

92) The ventral rami of spinal nerves T2-T12 are known as _____ nerves.

Answer: intercostal

93) The phrenic nerve arises from the _____ plexus.

Answer: cervical

94) The plexus that passes superior to the first rib posterior to the clavicle is the _____ plexus.

Answer: brachial

95) The largest nerve in the body is the _____.

Answer: sciatic

ESSAY. Write your answer in the space provided or on a separate sheet of paper.

96) Draw and describe the cross-sectional anatomy of the spinal cord.

Answer: The spinal cord is slightly flattened in its anterior-posterior dimension. It has two grooves—anterior median fissure and shallower posterior median sulcus. The central canal contains CSF. The gray commissure surrounds the central canal and contains gray horns (anterior, posterior, lateral), together forming an "H" arrangement. White matter surrounds the gray and is subdivided into columns (anterior, posterior, lateral). The anterior white commissure is anterior to the gray commissure.

97) Describe the gross external anatomy of the spinal cord.

Answer: The spinal cord is roughly cylindrical but slightly flattened anterior/posterior. It extends from the medulla to the superior border of L2 (16"–18") and is approximately 2 cm in diameter. There is a cervical enlargement from C4–T1 and a lumbar enlargement from T9–T12. The conus medullaris is the tapered inferior end. The filum terminale is an extension of the pia mater that anchors the cord to the coccyx. The cauda equina is the roots of spinal nerves angling inferiorly in the vertebral canal from the end of the spinal cord. 31 pairs of spinal nerves leave the cord at regular intervals.

98) Identify the components of a spinal reflex arc, and describe the function of each.

Answer:
1. Receptor responds to specific changes in environment by producing graded potential.
2. Sensory neuron conducts impulse from receptor to integrating center in gray matter of spinal cord.
3. Integrating center site of synapse between sensory and other neurons; decision-making area in gray matter of spinal cord.
4. Motor neuron conducts impulse from integrating center to effector.
5. Effector responds to motor nerve impulse; either muscle or gland.

99) Name and describe the locations of the four major plexuses of spinal nerves. Name a major nerve arising from each.

Answer:
1. cervical—alongside C1–C4—phrenic.
2. brachial—inferior and lateral to C4-T1 and superior to rib posterior to clavicle—axillary, ulnar, radial, median.
3. lumbar—lateral to L1-L4 passing obliquely posterior to psoas major and anterior to quadratus lumborum—femoral.
4. sacral—anterior to sacrum—sciatic.

100) George is 80 years old and has just developed painful blisters that seem to start abruptly at the anterior midline, circle around his waist on his right side, and stop equally abruptly on the posterior midline. What is probably wrong with George, and what accounts for the pattern of lesions?

Answer: George has shingles (herpes zoster). He probably had chickenpox as a child and some viruses stayed latent in his posterior root ganglia in his thoracic region. With age, his immunity to the virus has fallen off, and the virus has reactivated to travel via fast axonal transport to the skin surface. Lesions only appear on areas of skin served by the affected nerves (dermatomes).

CHAPTER 14   The Brain and Cranial Nerves

MULTIPLE CHOICE. Choose the one alternative that best completes the statement or answers the question.

1) The cavities within the brain are called
   A) sulci.
   B) choroid plexuses.
   C) nuclei.
   D) ventricles.
   E) commissures.

   Answer: D

2) Fine control of body coordination and balance is a function of the
   A) cerebellum.
   B) hypothalamus.
   C) thalamus.
   D) pituitary gland.
   E) reticular activating system.

   Answer: A

3) The falx cerebri is
   A) the deep groove between the precentral and postcentral gyri.
   B) the lobe of the cerebrum not visible on the surface.
   C) an extension of the dura mater separating the cerebral hemispheres.
   D) an extension of the dura mater separating the cerebellar hemispheres.
   E) an extension of the dura mater separating the cerebrum from the cerebellum.

   Answer: C

4) The brain requires such a large percentage of the body's total blood flow because
   A) it is such a large percentage of the body's total weight.
   B) its cells are always dividing to replace old cells.
   C) no glucose, the brain's exclusive energy source, is stored.
   D) it needs the heat to supply an alternative energy source.
   E) the movement of the blood stimulates the growth of the neuronal cytoplasmic extensions.

   Answer: C

5) Paired masses of gray matter within the midbrain that are rich in dopamine and are involved in maintenance of muscle tone are the

A) corpora quadrigemina.

B) basal ganglia.

C) mammillary bodies.

D) substantia nigra.

E) supraoptic nuclei.

Answer: D

6) The blood-brain barrier does not prevent passage of

A) any hormones.

B) most bacteria.

C) vitamins.

D) red blood cells.

E) lipid-soluble substances.

Answer: E

7) The type of neuroglia that wrap around capillaries in the blood-brain barrier are the

A) astrocytes.

B) Schwann cells.

C) oligodendrocytes.

D) microglia.

E) ependymal cells.

Answer: A

8) The main relay center for conducting information between the spinal cord and the cerebrum is the

A) thalamus.

B) insula.

C) corpus callosum.

D) cerebellar peduncles.

E) tentorium cerebelli.

Answer: A

9) The brain stem is made up of the

A) cerebellum, pons, and hypothalamus.

B) medulla oblongata, thalamus, and midbrain.

C) medulla oblongata, hypothalamus, and pons.

D) medulla oblongata, pons, and midbrain.

E) midbrain, hypothalamus, and thalamus.

Answer: D

10) Which of the following is NOT TRUE for cerebrospinal fluid?
    A) It acts as a shock absorber for the brain.
    B) It may contain white blood cells.
    C) It is a medium for exchange of nutrients and wastes.
    D) It acts as a neurotransmitter in the brain.
    E) It is produced by filtration and secretion in choroid plexuses.

    Answer: D

11) Superior to the hypothalamus and between the halves of the thalamus is the
    A) subthalamus.
    B) third ventricle.
    C) fourth ventricle.
    D) superior sagittal sinus.
    E) midbrain.

    Answer: B

12) The function of a choroid plexus is to
    A) receive sensations from the viscera.
    B) send motor impulses to the diaphragm.
    C) produce cerebrospinal fluid.
    D) reabsorb cerebrospinal fluid.
    E) transmit impulses from one cerebral hemisphere to the other.

    Answer: C

13) The function of arachnoid villi is to
    A) reabsorb cerebrospinal fluid.
    B) produce cerebrospinal fluid.
    C) hold the meninges onto the brain.
    D) provide nourishment for neurons in the CNS.
    E) conduct impulses from one cerebral hemisphere to the other.

    Answer: A

14) The blood-CSF barrier is composed of
    A) astrocytes.
    B) oligodendrocytes.
    C) myelin.
    D) dense irregular connective tissue.
    E) ependymal cells.

    Answer: E

15) The median and lateral apertures are the connection between the
    A) right and left cerebral hemispheres.
    B) right and left cerebellar hemispheres.
    C) fourth ventricle and subarachnoid space.
    D) third and fourth ventricles.
    E) lateral ventricles and third ventricle.

    Answer: D

16) If a blockage occurred in the interventricular foramina, what would be the likely result?
    A) hydrocephalus.
    B) anosmia.
    C) blindness.
    D) deafness.
    E) encephalitis.

    Answer: A

17) The pneumotaxic and apneustic centers are located in the
    A) medulla oblongata.
    B) pons.
    C) cerebellum.
    D) hypothalamus.
    E) cerebral cortex.

    Answer: B

18) Neurons from proprioceptors relay messages to the cerebellum by way of
    A) inferior olivary nucleus.
    B) inferior cerebellar peduncles.
    C) middle cerebellar peduncles.
    D) lateral geniculate nucleus.
    E) All are correct.

    Answer: A

19) White fibers that transmit impulses between corresponding gyri in opposite cerebral hemispheres are called
    A) association fibers.
    B) projection fibers.
    C) commissural fibers.
    D) ganglia.
    E) choroid plexuses.

    Answer: C

20) Damage to the cerebellum would result in
   A) loss of memory.
   B) uncoordinated movement.
   C) inability to dream.
   D) altered pituitary function.
   E) uncontrollable body temperature.

   Answer: B

21) The hippocampus is most important for
   A) conversion of short-term to long-term memory.
   B) production of cerebrospinal fluid.
   C) maintenance of posture.
   D) setting the basic pattern of breathing.
   E) controlling blood pressure.

   Answer: A

22) The reason hydrocephalus is so dangerous is that
   A) too many toxic products flood the brain.
   B) excess cerebrospinal fluid puts pressure on neurons, damaging them.
   C) the brain dehydrates.
   D) bacteria can grow more easily in accumulated fluid.
   E) it causes excessive release of hormones from the hypothalamus.

   Answer: B

23) Between the foramen magnum and the pons is the
   A) pituitary gland.
   B) hypothalamus.
   C) cerebellum.
   D) medulla oblongata.
   E) midbrain.

   Answer: D

24) The reticular activating system functions to
   A) maintain consciousness and awakefulness.
   B) maintain respiration.
   C) control blood pressure.
   D) maintain emotional patterns.
   E) All are correct.

   Answer: A

25) Most all sensory information synapses in the _____ and is relayed to the cerebral cortex.
   A) medulla oblongata
   B) pons
   C) basal ganglia
   D) thalamus
   E) hypothalamus

   Answer: D

26) The left side of the cerebrum controls skeletal muscles on the right side of the body because motor neurons cross from left to right in the
   A) precentral gyrus.
   B) cerebellum.
   C) medulla oblongata.
   D) hypothalamus.
   E) thalamus.

   Answer: C

27) The pneumotaxic and apneustic areas help control
   A) heart rate.
   B) blood pressure.
   C) breathing.
   D) skeletal movements.
   E) extracellular fluid volume.

   Answer: C

28) The cerebral peduncles are located in the
   A) forebrain.
   B) midbrain.
   C) hindbrain.
   D) All are correct.

   Answer: B

29) A sudden noise occurs behind you and you turn around in response. This is a reflex, which is the responsibility of the
   A) substantia nigra.
   B) medial lemniscus.
   C) superior colliculi.
   D) inferior colliculi.
   E) basal ganglia.

   Answer: D

30) Cerebellar peduncles attach the cerebellum to the
   A) medulla oblongata.
   B) pons.
   C) thalamus.
   D) hypothalamus.
   E) All are correct.

   Answer: B

31) The flocculonodular lobe of the cerebellum is concerned with the
   A) regulation of blood pressure.
   B) sense of smell.
   C) regulation of the pineal gland.
   D) regulation of breathing.
   E) sense of equilibrium.

   Answer: E

32) The lateral geniculate nucleus relays impulses from the retina to the cortex. This nucleus is contained in the
   A) hypothalamus.
   B) thalamus.
   C) cerebellum.
   D) basal ganglia.
   E) All are correct.

   Answer: B

33) The following are under the control of the hypothalamus EXCEPT
   A) releasing hormones for the anterior pituitary.
   B) autonomic nervous system.
   C) body temperature.
   D) production of melatonin.
   E) regulation of emotions.

   Answer: D

34) The satiety center is located in the
   A) thalamus.
   B) hypothalamus.
   C) epithalamus.
   D) basal ganglia.
   E) All are correct.

   Answer: B

35) Basal ganglia contain the following ganglia EXCEPT
    A) globus pallidus.
    B) putamen.
    C) amygdala.
    D) caudate nucleus.
    E) All answers are part of the basal ganglia.

    Answer: C

36) The limbic system includes the following parts EXCEPT
    A) cingulate gyrus.
    B) hippocampus.
    C) amygdala.
    D) putamen.
    E) All answers are part of the limbic system

    Answer: D

37) Releasing hormones that control the anterior pituitary gland are produced by the
    A) pineal gland.
    B) thalamus.
    C) hypothalamus.
    D) medulla oblongata.
    E) corpus callosum.

    Answer: C

38) Patterns of sleep are established by the
    A) suprachiasmatic nucleus of the hypothalamus.
    B) substantia nigra of the midbrain.
    C) corpora quadrigemina of the midbrain.
    D) decussation of pyramids in the medulla.
    E) arbor vitae of the cerebellum.

    Answer: A

39) The cerebral hemispheres are connected internally by the
    A) intermediate mass.
    B) basal ganglia.
    C) corpus callosum.
    D) arachnoid villi.
    E) habenular nuclei.

    Answer: C

40) The postcentral cortex is the
   A) primary somatosensory area.
   B) primary motor area.
   C) somatosensory association area.
   D) primary auditory area.
   E) All are correct.

   Answer: A

41) The primary motor area of the cerebral cortex is located in the
   A) precentral gyrus.
   B) postcentral gyrus.
   C) temporal lobe.
   D) occipital lobe.
   E) insula.

   Answer: A

42) The primary visual area and visual association area of the cerebral cortex are both located in the
   A) frontal lobe.
   B) temporal lobe.
   C) parietal lobe.
   D) insula.
   E) occipital lobe.

   Answer: E

43) The somatosensory cortex includes the following areas EXCEPT
   A) touch.
   B) proprioception.
   C) pain.
   D) thermal.
   E) All answers are correct.

   Answer: E

44) Damage to the cribriform plate of the ethmoid bone would most likely result in loss of
   A) vision.
   B) sensations to the side of the face.
   C) the sense of smell.
   D) the ability to speak.
   E) equilibrium.

   Answer: C

45) Speaking and understanding language are complex activities that are located in the
   A) Broca's area.
   B) Wernicke's area.
   C) common integrative area.
   D) premotor area.
   E) All answers are correct.

   Answer: A

46) Injury to the speech area, auditory association area and other language areas results in
   A) aphasia.
   B) agnosia.
   C) apraxia.
   D) dementia.
   E) delirium.

   Answer: A

47) During periods of sensory input and mental activity, the following brain waves could be seen on an EEG:
   A) alpha waves.
   B) beta waves.
   C) theta waves.
   D) delta waves.
   E) All are correct.

   Answer: B

48) A cerebral vascular accident is also referred to as a
   A) coronary.
   B) heart attack.
   C) stroke.
   D) dementia.
   E) All answers are correct.

   Answer: C

49) Temporary cerebral dysfunction including dizziness, weakness, etc is referred to as
   A) cerebral vascular accident.
   B) stroke.
   C) transcient ischemic attack.
   D) Alzheimer's disease.
   E) All answers are correct.

   Answer: C

50) At autopsy, the brains of Alzheimer's disease will show the following:

A) loss of neurons that liberate acetylcholine.

B) beta-amyloid plaques.

C) neurofibrillary tangles.

D) All answers are correct.

Answer: D

MATCHING. Choose the item in column 2 that best matches each item in column 1.

51) Outermost layer of the meninges.

52) Innermost layer of the meninges.

53) Cerebral spinal fluid located here.

54) Middle layer of the meninges.

A. Pia mater membrane

B. Subarachnoid space

C. Arachnoid membrane

E. Dura mater membrane

Answers: 51) D. 52) A. 53) B. 54) C.

55) Passes through the midbrain.

56) Passes from ventricles to third ventricle.

57) Exits fourth ventricle into subarachnoid space.

58) Passes from third ventricle to fourth ventricle.

A. Aqueduct of Sylvius

B. Aperture of Magendie

C. Foramina of Monro

D. Aperatures of Luschka

E. Both B and D.

Answers: 55) A. 56) C. 57) E. 58) A.

59) Optic nerve.

60) Olfactory nerve.

61) Taste.

62) Hearing.

A. Cranial Nerve I

B. Cranial Nerve II

C. Cranial Nerve VIII

D. Cranial Nerve IX

E. Cranial Nerve X

Answers: 59) B. 60) A. 61) D. 62) C.

63) Oculomotor nerve.

64) Originates in the pons.

65) Control secretion of saliva.

66) All participate in the autonomic nervous system.

A. Cranial Nerve III

B. Cranial Nerve VII

C. Cranial Nerve X

D. Cranial Nerve IX

E. All are correct.

Answers: 63) A. 64) B. 65) D. 66) E.

67) Primary respiratory centers.

68) Associated with cerebellar peduncles.

69) Relays most sensory input.

70) Regulates posture and balance.

A) Midbrain

B) Cerebellum

C) Thalamus

D) Medulla oblongata

E) Pons

Answers: 67) D. 68) E. 69) C. 70) B.

SHORT ANSWER. Write the word or phrase that best completes each statement or answers the question.

71) The cranial meninges include the outer _____, the middle _____, and the inner _____.

Answer: dura mater; arachnoid; pia mater

72) The large, dural venous sinus extending over the top of the brain is the _____.

Answer: superior sagittal sinus

73) The neuropeptides produced by the hypothalamus that are released as hormones from the posterior pituitary gland are _____ and _____.

Answer: oxytocin; antidiuretic hormone

74) Cranial nerve V is the _____ nerve.

Answer: trigeminal

75) The nuclei in the medulla oblongata that adjust the basic rhythm of breathing are called the _____.

Answer: medullary rhythmicity center

76) Cerebrospinal fluid flows from the fourth ventricle through the lateral and median apertures into the _____.

Answer: subarachnoid space

77) The superior sagittal sinus contains _____.

Answer: venous blood

78) The superior and inferior colliculi are the rounded elevations that make up the _____ of the midbrain.

Answer: corpora quadrigemina

79) The _____ is responsible for maintaining consciousness and for awakening from sleep.

Answer: reticular activating center

80) The superior cerebellar peduncles connect the _____ and the _____.

Answer: cerebellum; midbrain

81) The ventricles of the brain normally are filled with _____.

Answer: cerebrospinal fluid

82) Networks of capillaries involved in the production of cerebrospinal fluid are called _____.

Answer: choroid plexuses

83) The main brain region that regulates posture and balance is the _____.

Answer: cerebellum

84) Small brain regions in the walls of the third and fourth ventricles that can monitor chemical changes in blood because they lack a blood-brain barrier are called the _____.

Answer: circumventricular organs

85) Raising osmotic pressure in the extracellular fluid stimulates the thirst center located in the _____.

Answer: hypothalamus

86) Control of the ANS is an important function of the area of the diencephalon called the _____.

Answer: hypothalamus

87) The brain stem consists of the _____, the _____, and the _____.

Answer: medulla oblongata; pons; midbrain

88) The areas of the hypothalamus that regulate eating behavior are the _____ center, which is responsible for hunger sensations, and the _____ center that is stimulated when enough food has been ingested.

Answer: feeding; satiety

89) The _____ are part of the epithalamus involved in olfaction.

Answer: habenular nuclei

90) The red nucleus is a region of the _____ that is involved in coordinating muscular movements.

Answer: midbrain

91) The pineal gland secretes the hormone _____ that functions to _____.

Answer: melatonin; promote sleepiness (and help set the biological clock)

92) The upfolds of cerebral tissue are known as _____ or _____.

Answer: gyri; convolutions

93) The largest portion of the diencephalon is the _____.

Answer: thalamus

94) The infundibulum is a stalklike structure that attaches the _____ to the _____.

Answer: pituitary; hypothalamus

95) The postcentral gyrus of the cerebrum contains the primary _____ area.

Answer: somatosensory

ESSAY. Write your answer in the space provided or on a separate sheet of paper.

96) Describe the structure and function of the blood-brain barrier.

Answer: The barrier is formed by capillaries whose endothelial cells have tight junctions and a continuous basement membrane. Astrocytes also press against the capillaries to help control what can leave the blood and enter the brain. Small molecules and lipid-soluble substances pass easily; water-soluble substances may pass via carrier. Large molecules typically do not pass at all.

97) Explain why a crushing injury to the occipital bone is often fatal.

Answer: The crushing of the bone also crushes the brainstem, particularly the medulla oblongata. Damage to important nuclei regulating vital functions, such as respiration, heart rate and force of contraction, and diameter of blood vessels, may result in death.

98) Identify and describe the roles of the components of the brain involved in coordination of movement.

Answer: The following areas should be included: cerebellum, substantia nigra and red nucleus of midbrain, basal ganglia of cerebrum, subthalamus, and ventral nuclei of thalamus. The premotor area should be included as a motor association area.

99) Describe the structural and functional relationship between the hypothalamus and the pituitary gland.

Answer: The hypothalamus releases regulatory hormones into the capillary networks in the median eminence to increase or decrease hormone production and secretion from the anterior pituitary. Axons from the paraventricular and supraoptic nuclei extend through the infundibulum to the posterior pituitary. Their cell bodies produce oxytocin or antidiuretic hormone, which is stored and released from the axons that form the posterior pituitary.

100) Identify and describe the anatomical origins of those cranial nerves with exclusively or mainly sensory functions.

Answer:
1. Cranial nerve I (olfactory)—smell—nasal mucosa.
2. Cranial nerve II (optic)—vision—retina of eye.
3. Cranial nerve VIII (vestibulocochlear)—hearing and equilibrium—spiral organ of ear for hearing, and semicircular canals, utricle, and saccule of ear for equilibrium.

CHAPTER 15   Sensory, Motor, and Integrative Systems

MULTIPLE CHOICE. Choose the one alternative that best completes the statement or answers the question.

1)   The distinct quality that makes one sensation different from others is called its
   A) perception.
   B) receptive field.
   C) receptor potential.
   D) generator potential.
   E) modality.

   Answer: E

2)   Conscious awareness of external or internal stimuli defines the
   A) autonomic nervous system.
   B) motor nervous system.
   C) sensory nervous system.
   D) peripheral nervous system.

   Answer: C

3)   A sensory neuron's receptive field is the
   A) range of stimuli within a modality to which the neuron can respond.
   B) area of skin that is served by a particular sensory neuron.
   C) part of the neuron that is able to respond to the sensation.
   D) range of strengths of possible generator potentials for a particular neuron.
   E) all the different modalities to which a sensory neuron can respond.

   Answer: C

4)   Light is an appropriate stimulus for
   A) thermoreceptors.
   B) mechanoreceptors.
   C) chemoreceptors.
   D) photoreceptors.
   E) Both A and D are correct.

   Answer: D

5) The conscious awareness and interpretation of sensations is called
    A) modality.
    B) transduction.
    C) reception.
    D) perception.
    E) conduction.

    Answer: D

6) Precise location and identification of specific sensations occur in the
    A) spinal cord.
    B) thalamus.
    C) brain stem.
    D) cerebral cortex.
    E) All of the above.

    Answer: D

7) Sensory endings located in muscles, tendons, joints, and the inner ear are called
    A) exteroceptors.
    B) interoceptors.
    C) proprioceptors.
    D) chemoceptors.
    E) photoceptors.

    Answer: C

8) ALL of the following are considered general senses EXCEPT
    A) smell.
    B) vibration.
    C) pain.
    D) thermal sensations.
    E) joint and muscle position sense.

    Answer: A

9) Which of the following is NOT considered a special sense?
    A) smell
    B) vision
    C) pain
    D) taste
    E) equilibrium

    Answer: C

10) A stimulus that elicits a receptor potential causes

   A) direct release of neurotransmitter via exocytosis from synaptic vesicles.

   B) conduction of a nerve impulse along a first-order sensory nerve fiber.

   C) initiation of an action potential in the receptor.

   D) a subthreshold generator potential.

   E) reuptake of neurotransmitter from the extracellular fluid.

   Answer: A

11) Neurons that transmit impulses from the receptors to the central nervous system are called

   A) motor neurons.

   B) association neurons.

   C) bipolar neurons.

   D) sensory neurons.

   E) efferent neurons.

   Answer: D

12) A generator potential would be produced by ALL of the following EXCEPT

   A) Pacinian corpuscles.

   B) Merkel discs.

   C) tendon organs.

   D) Meissner corpuscles.

   E) taste buds.

   Answer: E

13) Exteroreceptors include receptors for ALL of the following EXCEPT

   A) vision.

   B) joint position.

   C) taste.

   D) smell.

   E) touch.

   Answer: B

14) Free nerve endings are

   A) bare dendrites.

   B) bare axons.

   C) encapsulated nerve endings.

   D) nonencapsulated nerve endings.

   E) All answers are correct.

   Answer: A

15) ALL of the following would be sensed by mechanoreceptors EXCEPT
   A) hearing.
   B) pressure.
   C) touch.
   D) equilibrium.
   E) light.

   Answer: E

16) Most all sensory impulses synapse in what structure on their way to the cerebral cortex?
   A) basal ganglia.
   B) corpus striatum.
   C) hypothalamus.
   D) mammillary bodies.
   E) thalamus.

   Answer: E

17) During a maintained, constant stimulus, a generator potential or a receptor potential decreases in amplitude in a phenomenon known as
   A) adaptation.
   B) transduction.
   C) propagation.
   D) perception.
   E) integration.

   Answer: A

18) Tactile sensations other than itch and tickle are detected by
   A) free nerve endings.
   B) encapsulated mechanoreceptors.
   C) special senses.
   D) encapsulated proprioceptors.
   E) nociceptors.

   Answer: B

19) Receptors for discriminative touch that are located in the stratum basale of the epidermis are the
   A) Meissner corpuscles.
   B) Pacinian corpuscles.
   C) Ruffini corpuscles.
   D) Merkel discs.
   E) intrafusal muscle fibers.

   Answer: D

20) Receptors for pressure that are widely distributed in the subcutaneous tissue and the submucosal tissues are the
   A) Meissner corpuscles.
   B) Pacinian corpuscles.
   C) Ruffini corpuscles.
   D) Merkel discs.
   E) intrafusal muscle fibers.

   Answer: B

21) Cold receptors are located in the
   A) subcutaneous layer.
   B) reticular layer of the dermis.
   C) stratum basale of the epidermis.
   D) dermal papillae.
   E) hair follicles.

   Answer: C

22) Intrafusal muscle fibers contract when stimulated by
   A) alpha motor neurons.
   B) upper motor neurons.
   C) gamma motor neurons.
   D) third-order neurons.
   E) interneurons.

   Answer: C

23) The brain sets an overall level of muscle tone by adjusting
   A) the angles at joints.
   B) the tension on a tendon.
   C) the amount of actin and myosin in muscle fibers.
   D) the conduction speed along type Ia sensory fibers.
   E) how vigorously a muscle spindle responds to stretching.

   Answer: E

24) The afferent part of stretch reflexes is provided by
   A) muscle spindles.
   B) tendon organs.
   C) joint kinesthetic receptors.
   D) lamellated corpuscles.
   E) type II cutaneous mechanoreceptors.

   Answer: A

25) Nociceptors are receptors for
    A) pain.
    B) phantom limb sensation.
    C) thermal sensation.
    D) hearing.
    E) equilibrium.

    Answer: A

26) An afferent neuron is a
    A) motor neuron.
    B) sensory neuron.
    C) neuromyojunction.
    D) All autonomic neurons.
    E) All are correct.

    Answer: B

27) Second-order neurons conduct nerve impulses from
    A) the PNS to the CNS.
    B) the spinal cord to the brain stem.
    C) the spinal cord and brain stem to the thalamus.
    D) the thalamus to the postcentral gyrus of the cerebral cortex.
    E) upper motor neurons to lower motor neurons.

    Answer: C

28) The primary somatosensory area of the cerebral cortex is located in the
    A) thalamus.
    B) occipital lobe.
    C) precentral gyrus.
    D) postcentral gyrus.
    E) hippocampus.

    Answer: D

29) Third-order sensory neurons in the posterior column-medial lemniscus pathway extend from the
    A) skin to posterior root ganglia.
    B) posterior root ganglia to the posterior gray horn of the spinal cord.
    C) spinal cord to the medulla oblongata.
    D) medulla oblongata to the thalamus.
    E) thalamus to the somatosensory area of the cerebral cortex.

    Answer: E

30) Injury to lower motor neurons results in
   A) kinesthesia.
   B) spastic paralysis.
   C) flaccid paralysis.
   D) coma.
   E) long-term potentiation.

   Answer: C

31) ALL of the following are indirect (extrapyramidal) motor tracts EXCEPT the
   A) rubrospinal tract.
   B) tectospinal tract.
   C) vestibulospinal tract.
   D) corticobulbar tract.
   E) lateroreticulospinal tract.

   Answer: D

32) Long-term potentiation is a phenomenon that occurs in
   A) formation of short-term memories.
   B) transition from NREM to REM sleep.
   C) conversion of short-term to long-term memory.
   D) destruction of neurons by the rubella virus.
   E) recovery of reflex activity following spinal shock.

   Answer: C

33) The cell bodies of first-order neurons in the posterior column-medial lemniscus pathway to the cortex are located in the
   A) posterior root ganglia of spinal nerves.
   B) medulla.
   C) medial lemniscus.
   D) anterior gray horns of the spinal cord.
   E) precentral gyrus.

   Answer: A

34) The medial lemniscus is a projection tract of second-order neurons extending from the
   A) posterior root ganglia to the medulla.
   B) proprioceptors to the anterior gray horns of the spinal cord.
   C) thalamus to the postcentral gyrus of the cerebral cortex.
   D) upper motor neurons to the lower motor neurons.
   E) medulla to the thalamus.

   Answer: E

35) About 90% of the neurons in the corticospinal tract cross in the
   A) pons.
   B) cerebrum.
   C) cerebellum.
   D) medulla oblongata.
   E) spinal cord.

   Answer: D

36) The pyramidal pathway is also called the
   A) spinothalamic.
   B) corticospinal.
   C) rubrospinal.
   D) vestibulospinal.
   E) All answers are correct.

   Answer: B

37) Direct motor pathways originate in the
   A) pre-central cortex.
   B) post-central cortex.
   C) pre-motor cortex.
   D) All answers are correct.

   Answer: A

38) Basal ganglia include the following:
   A) caudate nucleus.
   B) putamen.
   C) substantia nigra.
   D) globus pallidus.
   E) All answers are correct.

   Answer: E

39) Basal ganglia are supplied the neurotransmitter DOPAmine from
   A) putamen.
   B) substantia nigra.
   C) thalamus.
   D) hypothalamus.
   E) All answers are correct.

   Answer: B

40) Tendon organs are located
   A) interspersed within skeletal muscle fibers.
   B) at the junction of tendon and muscle.
   C) at the junction of tendon and bone.
   D) within the articular capsule.
   E) in the cerebellum.

Answer: B

41) Joint kinesthetic receptors are located
   A) interspersed within skeletal muscle fibers.
   B) at the junction of tendon and muscle.
   C) at the junction of tendon and bone.
   D) within the articular capsule.
   E) in the cerebellum.

Answer: D

42) Cell bodies for upper motor neurons are located in the
   A) precentral and postcentral gyri of the cerebral cortex.
   B) cerebellum.
   C) thalamus.
   D) anterior gray horns of the spinal cord.
   E) connective tissues surrounding skeletal muscles.

Answer: A

43) Lower motor neurons whose cell bodies are in nuclei in the brain stem stimulate
   A) upper motor neurons.
   B) interneurons.
   C) movements of the face and head.
   D) movements of limbs.
   E) parts of the cerebellum.

Answer: C

44) The stage of deep sleep dominated by delta waves on an EEG, and during which events such as bed-wetting and sleepwalking occur is
   A) REM sleep.
   B) Stage 1 NREM sleep.
   C) Stage 2 NREM sleep.
   D) Stage 3 NREM sleep.
   E) Stage 4 NREM sleep.

Answer: E

45) The internal capsule of the cerebrum contains
   A) axons of upper motor neurons.
   B) cell bodies of lower motor neurons.
   C) axons of second-order sensory neurons.
   D) axons of lower motor neurons.
   E) cell bodies of first-order sensory neurons.

   Answer: A

46) Parkinson's disease is due to the breakdown of the
   A) thalamus.
   B) hypothalamus.
   C) substantia nigra.
   D) precentral cortex.
   E) All answers are correct.

   Answer: C

47) Skilled movements of the limbs, hands and feet are controlled by motor neurons of the
   A) posterior columns.
   B) anterior spinothalamic tract.
   C) corticobulbar tract.
   D) tectospinal tract.
   E) lateral corticospinal tract.

   Answer: E

48) Axons passing through the cerebral peduncles and terminating in the nine pairs of cranial nerves in the pons and medulla belong to the
   A) posterior columns.
   B) anterior spinothalamic tract.
   C) corticobulbar tract.
   D) tectospinal tract.
   E) lateral corticospinal tract.

   Answer: C

49) The neurotransmitter for memory neuron synapses is
   A) DOPAmine.
   B) epinephrine.
   C) acetylcholine.
   D) norepinephrine.
   E) l-DOPA.

   Answer: C

50) Monitoring intention for movement and sending out corrective signals to coordinate movement are important functions of
   A) the primary somatosensory area.
   B) first-order neurons.
   C) third-order neurons.
   D) the thalamus.
   E) the cerebellum.

   Answer: E

MATCHING. Choose the item in column 2 that best matches each item in column 1.

51) Conveys motor impulses to muscles needed for maintaining balance in response to head movements.

A. Vestibulospinal tract

52) Indirect pathway that conveys motor impulses to muscles that move the head and eyes in response to visual stimuli.

B. Rubrospinal tract

53) Indirect pathway that conveys motor impulses to muscles involved d in precise movements of the limbs, hands and feet.

C. Tectospinal tract

54) Conveys motor impulses to muscles performing precise, voluntary movements of the head and neck.

D. Corticobulbar tract

E. Spinocerebar tract

   Answers: 51) A. 52) C. 53) B. 54) D.

55) Rapid adapting touch receptors.
56) Pressure receptors.
57) Slow adapting touch receptors.
58) Pain receptors.

A. Pineal corpuscles
B. Ruffini corpuscles
C. Meissner corpuscles
D. Pacinian corpuscles
E. Bare nerve endings

   Answers: 55) C. 56) D. 57) B. 58) E.

59) Somatosensory cortex.
60) Extrapyramidal motor cortex.
61) Corticospinal pathway begins with.
62) Memories are found here.

A. Precentral cortex
B. Postcentral cortex
C. Premotor cortex
D. Hippocamus cortex
E. All answers are correct

Answers: 59) B. 60) C. 61) A. 62) D.

63) Longterm memory potentiation.
64) Alzheimer's Disease is a deficiency in.
65) Substantia nigra.
66) Reticular activating system.

A. Acetylcholine
B. DOPAmine
C. Glutamate
D. Epinephrine
E. All are correct

Answers: 63) C. 64) A. 65) B. 66) D.

67) A motor disorder that causes loss of muscle control.
68) Difficulty in falling asleep.
69) State of deep unconsciousness.
70) Person stops breathing during sleep.

A. Coma
B. Insomnia
C. Narcolepsy
D. Sleep apnea
E. Cerebral palsy

Answers: 67) E. 68) B. 69) A. 70) D.

SHORT ANSWER. Write the word or phrase that best completes each statement or answers the question.

71) A sense organ transduces a stimulus into a _____ .

Answer: graded potential

72) The portion of a receptor that is capable of responding to stimulation is called the _____ .

Answer: receptive field

73) Receptors of discriminative touch located in the dermal papillae of hairless skin are called _____ .

Answer: corpuscles of touch (Meissner corpuscles)

74) The perception of body movements is called _____ .

Answer: kinesthesia

75) Long-term potentiation is a phenomenon believed to occur in a region of the cerebrum known as the _____ .

Answer: hippocampus

76) The special cells that make up a muscle spindle are called _____.

Answer: intrafusal muscle fibers

77) The ability to recognize by "feel" the size, shape, and texture of an object is called _____.

Answer: stereognosis

78) _____ is the conscious or unconscious awareness of external or internal stimuli.

Answer: Sensation

79) Each specific type of sensation is called a sensory _____.

Answer: modality

80) Muscle tone is regulated by the brain as it adjusts the level of sensitivity of receptors called _____ to the degree of muscle stretching.

Answer: muscle spindles

81) Receptors that provide information concerning body position and movement are classified as _____.

Answer: proprioceptors

82) _____ initiate reflexes that cause muscle relaxation to decrease muscle tension before damage occurs.

Answer: Tendon organs

83) The proprioceptors present in and around the articular capsules of synovial joints are called _____.

Answer: joint kinesthetic receptors

84) A decrease in sensitivity to a long-term stimulus is called _____.

Answer: adaptation

85) _____ neurons conduct impulses from the spinal cord and brain stem to the thalamus.

Answer: Second-order

86) A projection tract of sensory neurons that extends from the medulla to the thalamus is the _____.

Answer: medial lemniscus

87) Visceral pain that is perceived as localized in the skin served by the same segment of the spinal cord is called _____.

Answer: referred pain

88) The axons of first-order neurons transmitting impulses of conscious proprioception form the spinal tracts known as the _____.

Answer: posterior columns

89) Axons of second-order neurons in the anterolateral pathways synapse with third-order neurons in the _____.

Answer: thalamus

90) The region of the cerebral cortex that provides the major control for initiation of voluntary movements is the _____.

Answer: primary motor area (precentral gyrus)

91) _____ are the only neurons that carry information from the CNS to skeletal muscle fibers.

Answer: Lower motor neurons

92) Most upper motor neurons _____ (or cross over) to the contralateral side of the medulla.

Answer: decussate

93) Skilled movements of the limbs, hands, and feet are controlled by motor neurons of the _____ tracts.

Answer: lateral corticospinal

94) Awakening from sleep involves increased activity in fibers known as the _____ that project from the brain stem through the thalamus to the cerebral cortex.

Answer: reticular activating system

95) Damage to lower motor neurons produces _____ paralysis of muscles on the same side of the body.

Answer: flaccid

ESSAY. Write your answer in the space provided or on a separate sheet of paper.

96) Describe the possible relationships between sensory receptors and first-order sensory neurons.

Answer: Receptors for general senses could be free or encapsulated nerve endings of first-order sensory neurons in which stimuli elicit generator potentials. Receptors for the special senses consist of separate cells in which stimuli elicit receptor potentials that trigger exocytosis of neurotransmitter which diffuses across a synapse to initiate nerve impulses in a first-order neuron.

97) Compare short-term memory vs. long-term memory with regard to specific changes that are thought to occur in the brain.

Answer: Short-term memory may depend on forming new synapses and reverberating circuits. Long-term memory is thought to involve high-frequency stimulation within the hippocampus at glutamate synapses. Nitric oxide and other neurotransmitters may be involved. Neurons develop new presynaptic terminals, larger synaptic end bulbs, and more dendritic branches. Enhanced facilitation occurs. DNA and RNA are possibly involved.

98) Describe the role of the reticular activating system in sleep, arousal, and consciousness.

Answer: A variety of sensory stimuli feed into the RAS, which feeds into the thalamus and cerebral cortex to increase neuronal activity, causing arousal from sleep and maintaining consciousness. During periods of high ATP use, adenosine accumulates and binds to A1 receptors, inhibiting cholinergic neurons in the RAS, and inducing sleep.

99) You are sitting on a sunny Florida beach. Describe the anatomical structures that must be present and the physiological events that must occur in order for you to perceive the warmth of the sun.

Answer: Thermoreceptors must be present to sense the thermal stimuli. The stimulus must occur within the receptor's receptive field. The stimuli are transduced into graded potentials that eventually reach threshold to trigger nerve impulses in a first-order neuron, which transmits the information to the CNS via the neurons of the spinothalamic tract. The primary somatosensory area then must interpret the information provided by these neurons.

100) Two children were born in England who lacked nociceptors. What effects would this have on the lives of these children? Explain your answer.

Answer: Lack of pain sensations would require careful monitoring by parents and the children themselves for tissue damage (or the potential thereof . . .). The question might be expanded to include speculation about the social/behavioral effects of this condition.

Chapter 16   The Special Senses

MULTIPLE CHOICE. Choose the one alternative that best completes the statement or answers the question.

1) The sites of olfactory transduction are the olfactory hairs, which are
   A) cilia projecting from the dendrites of first-order neurons.
   B) microvilli on the axons of first-order neurons.
   C) columnar epithelial cells in the nasal mucosa.
   D) mucus-producing cells in the nasal epithelium.
   E) projection tracts between the thalamus and the temporal lobe.

   Answer: A

2) G proteins activated by binding of odorant molecules to olfactory receptors cause
   A) exocytosis of neurotransmitter.
   B) activation of adenylate cyclase.
   C) influx of calcium ions.
   D) bending of olfactory hairs.
   E) hyperpolarization of olfactory neurons.

   Answer: B

3) First-order olfactory neurons synapse with second-order neurons in the
   A) olfactory epithelium.
   B) cribriform plate.
   C) olfactory tract.
   D) temporal lobe.
   E) olfactory bulbs.

   Answer: E

4) Olfactory sensations reach the brain via
   A) cranial nerve I.
   B) cranial nerve II.
   C) cranial nerve VII.
   D) cranial nerve VIII.
   E) spinal nerves C1.

   Answer: A

5) The receptors on the tip of the tongue are most sensitive to

   A) sweet.

   B) sour.

   C) salty.

   D) bitter.

   E) None of these; there are no taste buds at the tip of the tongue.

   Answer: A

6) In order for a substance to be tasted, it must be

   A) a partially denatured protein.

   B) of a pH below 7.

   C) an ionic compound.

   D) dissolved in saliva.

   E) Both C and D are correct.

   Answer: D

7) Fungiform papillae contain

   A) gustatory cells.

   B) olfactory cells.

   C) lacrimal cells.

   D) salivary cells.

   E) All answers are correct.

   Answer: A

8) The primary gustatory area of the cerebrum is in the

   A) parietal lobe.

   B) frontal lobe.

   C) temporal lobe.

   D) occipital lobe.

   E) midbrain.

   Answer: A

9) Most impulses related to gustatory sensations arising on the tongue are conveyed to the brain via

   A) cranial nerve I.

   B) cranial nerve II.

   C) cranial nerve VII.

   D) cranial nerve VIII.

   E) spinal nerves C1.

   Answer: C

10) The lacrimal apparatus produces

   A) sweat.

   B) sebum.

   C) aqueous humor.

   D) tears.

   E) Both C and D are correct.

   Answer: D

11) Lacrimal fluid contains a protective bacterial enzyme called

   A) isoenzyme.

   B) lysozyme.

   C) isozyme.

   D) All answers are correct.

   Answer: B

12) Venous sinus (Canal of Schlemm) exits the eye through an opening in the

   A) ciliary body.

   B) ciliary muscle.

   C) sclera.

   D) retina.

   E) choroid.

   Answer: C

13) "Bloodshot eyes" are the result of dilation of blood vessels in the

   A) lens.

   B) cornea.

   C) sclera.

   D) conjunctiva.

   E) iris.

   Answer: D

14) The function of the ciliary processes is to

   A) alter the shape of the lens.

   B) prevent objects from falling into the eye.

   C) produce aqueous humor.

   D) change the diameter of the pupil.

   E) transduce light stimuli into nerve impulses.

   Answer: C

15) The lens is held in place by the
    A) choroid.
    B) iris.
    C) ciliary processes.
    D) vitreous body.
    E) suspensory ligaments.

    Answer: E

16) Intraocular pressure is produced mainly by
    A) aqueous humor.
    B) vitreous humor.
    C) constriction of the pupil.
    D) lacrimal secretions.
    E) photopigments.

    Answer: A

17) The spaces anterior to the lens are filled with
    A) aqueous humor.
    B) vitreous humor.
    C) photoreceptors.
    D) lacrimal gland secretions.
    E) air.

    Answer: A

18) Nutrients are provided to the posterior surface of the retina by blood vessels in the darkly pigmented portion of the vascular tunic known as the
    A) choroid.
    B) ciliary body.
    C) iris.
    D) sclera.
    E) macula lutea.

    Answer: A

19) The retina is held in place by the
    A) optic disc.
    B) vitreous body.
    C) ciliary muscle.
    D) bipolar neurons.
    E) iris.

    Answer: B

20) Photoreceptors are located in the
    A) choroid.
    B) cornea.
    C) iris.
    D) retina.
    E) Both C and D are correct.

    Answer: D

21) The blind spot of the retina is so-called because
    A) only cones are found there.
    B) only rods are found there.
    C) there are no rods or cones there.
    D) it is impossible for light rays to be focused on that spot.
    E) inhibitory neurotransmitters are released from the receptors there.

    Answer: C

22) The central fovea is
    A) where the optic nerve exits the back of the eye.
    B) the area of highest visual acuity on the retina.
    C) what secretes aqueous humor.
    D) where the retinal artery branches.
    E) where new photopigments are produced.

    Answer: B

23) For sharpest visual acuity, light rays must be refracted so that they
    A) change wavelength to fall within the visible range.
    B) stimulate constriction of the pupil.
    C) turn photopigments in the lens into their colorless form.
    D) hit the melanin in the choroid.
    E) fall directly on the central fovea.

    Answer: E

24) In the process of forming an image on the retina, convergence occurs to allow
    A) a change in shape of the lens.
    B) three-dimensional image formation.
    C) refraction of light rays.
    D) focusing of light through the center of the lens.
    E) production of new photoreceptors.

    Answer: B

25) The central fovea is a small depression in the center of the
   A) macula lutea of the retina.
   B) optic disc of the retina.
   C) lens.
   D) cornea.
   E) vitreous body.

   Answer: A

26) Which of the following occurs when trying to focus on a close object?
   A) relaxation of the ciliary muscle to flatten the lens
   B) relaxation of the ciliary muscle to make the lens more convex
   C) contraction of the ciliary muscle to flatten the lens
   D) contraction of the ciliary muscle to make the lens more convex
   E) contraction of the ciliary muscle to increase the diameter of the pupil

   Answer: D

27) The four refracting media of the eye, listed in the sequence in which they refract light, are
   1. vitreous body.
   2. lens.
   3. aqueous humor.
   4. cornea.
   A) 1, 2, 3, and 4
   B) 4, 1, 2, and 3.
   C) 4, 3, 2, and 1
   D) 2, 3, 4, and 1
   E) 3, 2, 1, and 4

   Answer: C

28) The lens of the eye thickens when the
   A) ciliary muscles relax.
   B) ciliary processes relax.
   C) ciliary muscles contract.
   D) suspensory ligaments pull on the lens capsule.

   Answer: C

29) Which of the following occurs in the hypermetropic eye?
    A) Light rays converge on the central fovea.
    B) Light rays converge in front of the central fovea.
    C) Light rays converge behind the central fovea.
    D) Intraocular pressure builds to damage neurons.
    E) The lens becomes cloudy.

    Answer: C

30) In the accommodation reflex for close-up vision, what adjustments are made?
    A) The ciliary muscles contract, the lens become more convex and the pupil constricts.
    B) The muscles of the ciliary body reflex, the lens becomes less convex and the sphincter of the pupil relaxes.
    C) The ciliary muscles contract, tightening the suspensory ligaments, the lens flattens, and the pupil becomes dark adapted.
    D) The ciliary muscles contract.
    E) The extrinsic muscles contract, the lens does not change, but the radial muscle relax.

    Answer: A

31) Which of the following is formed directly from vitamin A?
    A) rhodopsin
    B) lumirhodopsin
    C) metarhodopsin
    D) retinene
    E) scotopsin

    Answer: D

32) Suffering from Vitamin A deficiency leads to
    A) tunnel vision.
    B) glaucoma.
    C) night blindness.
    D) color blindness.
    E) astigmatism.

    Answer: C

33) Which one of the following is the correct sequence for the passage by a nerve impulse?

1. lateral geniculate bodies

2. optic tract

3. optic II nerve

4. optic chiasma

5. occipital lobe

A) 3, 4, 2, 1, 5.

B) 1, 2, 3, 4, 5.

C) 2, 3, 4, 1, 5.

D) 5, 1, 2, 4, 3.

E) 4, 3, 2, 5, 1.

Answer: A

34) The characteristic shared by photopigments in rods and cones is that they all

A) contain the same opsin molecule.

B) respond to the same wavelengths of light.

C) contain retinal as the light-absorbing molecule.

D) Both A and C are correct.

E) A, B, and C are all correct.

Answer: C

35) Which of the following lists the route of impulse transmission in the correct order?

A) photoreceptors, ganglion cells, bipolar cells, optic nerve, optic chiasm, optic tract

B) bipolar cells, ganglion cells, photoreceptors, optic nerve, optic tract, optic chiasm

C) photoreceptors, bipolar cells, ganglion cells, optic nerve, optic tract, optic chiasm

D) photoreceptors, bipolar cells, ganglion cells, optic nerve, optic chiasm, optic tract

E) bipolar cells, ganglion cells, photoreceptors, optic nerve, optic chiasm, optic tract

Answer: D

36) The optic nerve is made up of the

A) axons of the rods and cones.

B) axons of the bipolar cells.

C) axons of the ganglion cells.

D) dendrites of the bipolar cells.

E) dendrites of the ganglion cells.

Answer: C

37) What happens at the optic chiasm?
   A) Rods stimulate bipolar cells.
   B) The optic nerve exits the eye.
   C) Visual input is perceived.
   D) The optic nerves cross.
   E) Aqueous humor is reabsorbed.

   Answer: D

38) Severing the right optic tract would
   A) destroy all sight in the right eye.
   B) destroy all sight in the left eye.
   C) eliminate sight in the right nasal and left temporal visual fields.
   D) eliminate sight in the left nasal and right temporal visual fields.
   E) All of the answers are correct.

   Answer: D

39) The neurotransmitter in rods is
   A) acetylcholine.
   B) substance P.
   C) glutamate.
   D) dopamine.
   E) epinephrine.

   Answer: C

40) A nearsighted individual
   A) has shortened eyeball.
   B) uses biconcave lens for correction.
   C) has the image of an object focused behind the fovea.
   D) uses convex lens for correction.

   Answer: B

41) The auditory (Eustachian) tube connects the
   A) middle ear and inner ear.
   B) external ear and middle ear.
   C) middle ear and nasopharynx.
   D) cochlea and vestibule.
   E) inner ear and primary auditory area of the cerebrum.

   Answer: C

42) The ossicles are the major structures of the

A) external ear.

B) middle ear.

C) vestibule.

D) cochlea.

E) auditory regions of the cerebrum.

Answer: B

43) The middle ear is normally filled with

A) air.

B) blood.

C) endolymph.

D) perilymph.

E) cerumen.

Answer: A

44) The role of the stapedius muscle is to

A) stabilize the tympanic membrane.

B) connect the ossicles to the cochlea.

C) flex the hair cells of the spiral organ.

D) constrict the Eustachian tube.

E) limit vibration of the stapes.

Answer: E

45) Receptor potentials are produced when

A) the tympanic membrane moves the malleus.

B) the stapes pushes into the oval window.

C) perilymph moves in the scala vestibuli.

D) stereocilia bend against the tectorial membrane.

E) All of these generate receptor potentials.

Answer: D

46) First-order sensory neurons of the cochlear branch of the vestibulocochlear nerve terminate in the

A) thalamus.

B) occipital lobe.

C) basilar membrane.

D) medulla oblongata.

E) vestibular branch.

Answer: D

47) The primary function of the utricle and saccule is to

   A) transduce sound waves into generator potentials.

   B) monitor static equilibrium.

   C) monitor dynamic equilibrium.

   D) cause movements of the ossicles.

   E) produce endolymph in the cochlea.

   Answer: B

48) The receptors in the utricle and saccule are stimulated when

   A) endolymph flows over the spiral organ.

   B) the otoliths bend stereocilia in response to gravity.

   C) the tympanic membrane vibrates.

   D) perilymph bends hair cells.

   E) the stapes pushes into the oval window.

   Answer: B

49) Which one of the following sequences corectly traces the sound wave across the middle ear?

   1. tympanic membrane

   2. malleus

   3. incus

   4. stapes

   5. oval window

   A) 5, 4, 2, 3, 1.

   B) 1, 2, 3, 4, 5.

   C) 2, 4, 5, 1, 3.

   D) 3, 1, 4, 2, 5.

   E) 4, 5, 1, 2, 3.

   Answer: B

50) The endolymph filled_____lies between the perilymph filled _____and the _____separated by the _____and ____membranes respectively.

   A) scala vestibuli, scala tympani, scala ampularis, cochlear, vestibular

   B) road window, oval window, semicircular canals, tympanic tectorial

   C) scala tympani, cochlear duct, scala vestibuli, tympanic, basilar

   D) cochlear duct, scala vestibuli, scala tympani, basilar, vestibular

   E) cochlear duct, oval window, round window, tympanic, cochlear

   Answer: D

MATCHING. Choose the item in column 2 that best matches each item in column 1.

51) Ossicle attached to the internal surface of the eardrum.

52) Connects the middle ear with the nasopharynx.

53) Ossicle that fits into the oval window.

54) Separates external auditory canal from the middle ear.

A. Malleus

B. Incus

C. Stapes

D. Tympanic Membrane

E. Auditory tube

Answers: 51) A. 52) E. 53) C. 54) D.

55) Contains the utricle and the saccule.

56) Structure pulled by gravity over hair cells involved in static equilibrium.

57) Fluid that fills the body labyrinth.

58) Fluid that fills the membranous labyrinth.

A. Vestibule

B. Otolithic membrane

C. Spiral organ

D. Endolymph

E. Perilymph

Answers: 55) A. 56) B. 57) E. 58) D.

59) Membranous labyrinth in the vestibule consists of a sac called.

60) Cell bodies of sensory neurons.

61) End of each semicircular canal is an enlargement called.

62) Three bony ducts.

A. Semicircular canal

B. Ampulla

C. Utricle

D. Ampulla

E. Vestibular ganglia

Answers: 59) C. 60) E. 61) D. 62) A.

63) Inverted V-shaped row at the back of tongue.

64) Mushroom-shaped elevations over entire surface of tongue.

65) Threadlike structures over entire tongue.

66) Small trenches on the lateral margins of tongue.

A. Filiform papillae

B. Fungiform papillae

C. Vallate papillae

D. Foliate papillae

E. All answers are correct

Answers: 63) C. 64) B. 65) A. 66) D.

67)	The "white" of the eye.	A)	Cornea
68)	Avascular, transparent part of the fibrous tunic.	B)	Vitreous body
69)	Jellylike substance in posterior cavity that helps hold retina against the eyeball.	C)	Sclera
70)	Highly vascular, darkly pigmented portion of uvea.	D)	Aqueous humor
		E)	Choroid

Answers: 67) C. 68) A. 69) B. 70) E.

SHORT ANSWER. Write the word or phrase that best completes each statement of answers the question.

71) New olfactory receptors are produced by the division of cells within the olfactory epithelium called _____.

Answer: basal stem cells

72) The _____ have the primary role in maintenance of dynamic equilibrium.

Answer: semicircular ducts

73) Axons of cranial nerve I terminate in paired masses of gray matter inferior to the frontal lobes of the cerebrum known as the _____.

Answer: olfactory bulbs

74) The receptors for gustatory sensations are located in gustatory receptors known as _____.

Answer: taste buds

75) The four primary tastes are _____ , _____ , _____ , and _____.

Answer: sweet; sour; salty; bitter

76) The anatomical name for the eyelids is the _____.

Answer: palpebrae

77) The shape of the lens is altered for near or far vision by the _____.

Answer: ciliary muscle

78) The hole in the center of the iris is the _____.

Answer: pupil

79) Photoreceptors called _____ are most important for seeing shades of gray in dim light, while _____ provide color vision in bright light.

Answer: rods; cones

80) The mucous membrane lining the eyelids and covering part of the anterior surface of the eyeball is the _____.

Answer: conjunctiva

81) The dense connective tissue making up the white part of the fibrous tunic is called the _____.

Answer: sclera

82) The highly vascularized layer of the uvea that lines most of the internal surface of the sclera is the _____.

Answer: choroid

83) Aqueous humor is secreted by blood capillaries located in the _____ of the vascular tunic.

Answer: ciliary processes

84) In dim light, _____ neurons stimulate _____ muscles of the iris to contract, causing a(n) _____ in the diameter of the pupil.

Answer: sympathetic; radial; increase

85) The jellylike substance filling the posterior cavity of the eye is the _____.

Answer: vitreous body

86) Bending of light rays as they pass through different media is called _____.

Answer: refraction

87) The site on the retina where the optic nerve exits the eyeball is the _____.

Answer: optic disc

88) Reflection and scattering of light rays within the eyeball is prevented by the absorption of stray light rays by _____ in the choroid and pigment epithelium of the retina.

Answer: melanin

89) Aqueous humor drains from the anterior chamber back into the blood via the _____.

Answer: scleral venous sinus (canal of Schlemm)

90) The photopigment in rods is _____.

Answer: rhodopsin

91) The first point of refraction as light passes into the eye is the _____.

Answer: cornea

92) The external auditory canal and the middle ear are separated by the _____.

Answer: tympanic membrane

93) The auditory ossicles are the _____ , the _____ , and the _____ .

Answer: malleus; incus; stapes

94) A refraction abnormality in which either the cornea or the lens has an irregular curvature is called _____.

Answer: astigmatism

95) The light-absorbing portion of all visual photopigments is _____ , which is derived from vitamin _____.

Answer: retinal; A

ESSAY. Write your answer in the space provided or on a separate sheet of paper.

96) Describe the process of image formation on the retina.

Answer:
1) Refraction—bending of light as medium changes to focus light into central fovea.
2) Accommodation of lens for near/distance vision—shape of lens changed by ciliary muscle to make light focus on retina.
3) Constriction of pupil—ANS reflex to prevent scattering of light through edges of lens.
4) Convergence of eyes—to focus both eyes on same object and provide binocular (3D) vision.

Images are focused on the retina upside-down and mirror-image, and the brain then translates this information.

97) Explain the process by which smell sensations are sensed and perceived.

Answer: Odorant molecules dissolve in mucus secreted by the olfactory epithelium and bind to receptors, triggering a generator potential. In some cases the binding activates a G protein in the plasma membrane that activates adenylate cyclase which opens sodium ion channels. Axons of the receptors (first-order neurons) transmit impulses via cranial nerve I through the olfactory foramina of the cribriform plate and terminate in the olfactory bulbs, where they synapse with second-order neurons. These axons form the olfactory tracts, which transmit impulses to the olfactory area in the temporal lobe. Other important brain areas include the limbic system, the hypothalamus, and the orbitofrontal area.

98) Describe the anatomical features that provide protection to the delicate neural portion of the retina and that ensure the retina is kept flat against the back of the eye.

Answer: The bony orbit provides protection to most of the eye. The sclera makes the eyeball rigid and, by its toughness, prevents foreign objects from entering the eye. Aqueous fluid creates intraocular pressure to support the eye from inside, and the vitreous body holds the retina flush against the back of the eye.

99) Differentiate between static and dynamic equilibrium. Describe the structures and physiological mechanisms involved in receiving and transducing vestibular sensations.

Answer: Static equilibrium is the maintenance of body position relative to gravity. Hair cells in the maculae of the utricle and saccule bend as the otolithic membrane slides forward due to gravity. Receptor potentials are transmitted to cranial nerve VIII to the pons. Dynamic equilibrium is the maintenance of body position in response to movement. Endolymph flowing over hair cells in the cristae of semicircular ducts causes bending. Receptor potentials are passed to cranial nerve VIII to the pons.

100) Blindness can occur for many reasons. Using your knowledge of the structure of the eye and the processing of light stimuli, predict where some of the potential trouble spots might be. Explain your answers.

Answer: Examples could include: loss of transparency of the cornea or lens (light can't pass); detachment of the retina (loss of photoreceptive and conductive function); no photoreceptors or pigments (no transduction); damage to the optic nerve (no conduction); damage to the visual pathway or visual cortex (no conduction or translation).

CHAPTER 17   The Autonomic Nervous System

MULTIPLE CHOICE. Choose the one alternative that best completes the statement or answers the question.

1) To say that most organs served by the ANS have "dual innervation" means that
    A) these organs release either acetylcholine or norepinephrine when stimulated.
    B) it takes two postganglionic neurons to achieve the desired response.
    C) the organs are innervated by both sympathetic and parasympathetic neurons.
    D) the organs have both alpha and beta receptors.
    E) both a preganglionic and postganglionic neuron go to the organ.

    Answer: C

2) Which of the following is an example of an effector for a visceral efferent neuron?
    A) quadriceps femoris
    B) diaphragm
    C) extrinsic eye muscles
    D) smooth muscle in the wall of the small intestine
    E) All of the above are correct except the quadriceps femoris

    Answer: D

3) A small-diameter, myelinated type B neuron that terminates in an autonomic ganglion is a(n)
    A) autonomic sensory neuron.
    B) preganglionic neuron.
    C) postganglionic neuron.
    D) somatic motor neuron.
    E) adrenergic neuron.

    Answer: B

4) A small-diameter unmyelinated type C neuron that terminates in a visceral effector is a(n)
    A) autonomic sensory neuron.
    B) preganglionic neuron.
    C) postganglionic neuron.
    D) white ramus.
    E) splanchnic nerve.

    Answer: C

5) The lateral gray horns of the thoracic spinal cord is the location of
A) splanchnic nerves.
B) sympathetic chains.
C) synapses between preganglionic and postganglionic neurons.
D) cell bodies of some sympathetic preganglionic neurons.
E) cell bodies of some parasympathetic neurons.

Answer: D

6) The term *craniosacral outflow* refers to
A) release of acetylcholine from preganglionic neurons.
B) axons of parasympathetic postganglionic neurons.
C) axons of sympathetic postganglionic neurons.
D) axons of parasympathetic preganglionic neurons.
E) axons of sympathetic preganglionic neurons.

Answer: D

7) The lumbar region of the spinal cord is the location of cell bodies of
A) parasympathetic preganglionic fibers.
B) parasympathetic postganglionic fibers.
C) sympathetic preganglionic fibers.
D) sympathetic postganglionic fibers.
E) Both A and C are correct.

Answer: C

8) Thoracolumbar division is called the
A) sympathetic division.
B) parasympathetic division.
C) CNS sensory division.
D) CNS motor division.
E) All answers are correct.

Answer: A

9) The first of two motor neurons in any autonomic motor pathway is called
A) premotor neuron.
B) preganglionic neuron.
C) postmotor neuron.
D) postganglionic neuron.
E) All answers are correct.

Answer: B

10) The second neuron in the autonomic motor pathway is called
   A) premotor neuron.
   B) preganglionic neuron.
   C) postmotor neuron.
   D) postganglionic neuron.
   E) All answers are correct.

   Answer: D

11) Cell bodies of preganglionic neurons of the parasympathetic division are located in the nuclei of ALL of the following cranial nerves EXCEPT
   A) III.
   B) V.
   C) VII.
   D) IX.
   E) X.

   Answer: B

12) Sympathetic ganglia are also called
   A) sympathetic trunk ganglia.
   B) vertebral chain ganglia.
   C) paravertebral ganglia.
   D) All answers are correct.

   Answer: D

13) The sympathetic ganglia that lie close to large abdominal arteries are the
   A) prevertebral ganglia.
   B) paravertebral ganglia.
   C) terminal ganglia.
   D) sympathetic trunk ganglia.
   E) dorsal root ganglia.

   Answer: A

14) Autonomic plexuses include the
   A) cardiac plexus.
   B) pulmonary plexus.
   C) celiac plexus.
   D) renal plexus.
   E) All answers are correct.

   Answer: E

15) In which of the following would synapses between sympathetic preganglionic and postganglionic fibers occur?
    A) ciliary ganglion
    B) superior mesenteric ganglion
    C) celiac ganglion
    D) otic ganglion
    E) Both B and C are correct.

    Answer: E

16) Sympathetic chains are
    A) preganglionic fibers that pass through sympathetic trunk ganglia and extend to prevertebral ganglia.
    B) connections between sympathetic and parasympathetic ganglia.
    C) fibers extending from sympathetic trunk ganglia to effectors.
    D) the autonomic part of a spinal nerve.
    E) axon collaterals of sympathetic fibers extending between sympathetic trunk ganglia.

    Answer: E

17) Sympathetic preganglionic fibers that pass through sympathetic trunk ganglia to synapse in prevertebral ganglia are called
    A) white rami communicantes.
    B) gray rami communicantes.
    C) splanchnic nerves.
    D) sympathetic chains.
    E) anterior roots.

    Answer: C

18) The structure containing sympathetic postganglionic fibers that connect sympathetic trunk ganglia to spinal nerves is the
    A) sympathetic chain.
    B) anterior root.
    C) posterior root.
    D) white ramus communicans.
    E) gray ramus communicans.

    Answer: E

19) In the autonomic nervous system, all preganglionic fibers release the neurotransmitter
    A) acetylcholine.
    B) norepinephrine.
    C) serotonin.
    D) dopamine.
    E) epinephrine.

Answer: A

20) In the autonomic nervous system, most sympathetic postganglionic fibers release the neurotransmitter
    A) acetylcholine.
    B) norepinephrine.
    C) serotonin.
    D) dopamine.
    E) glutamate.

Answer: B

21) Most autonomic sensory neurons are associated with
    A) exteroreceptors.
    B) interoreceptors.
    C) proprioceptors.
    D) special senses.
    E) somatic efferent neurons.

Answer: B

22) An adrenergic neuron produces the neurotransmitter
    A) serotonin.
    B) GABA.
    C) norepinephrine.
    D) acetylcholine.
    E) glycine.

Answer: C

23) Which of the following responses is initiated by the sympathetic nervous system?
    A) decreased heart rate
    B) constriction of pupils
    C) splitting glycogen to glucose by the liver
    D) constriction of the bronchioles
    E) decreased blood pressure

Answer: C

24) Parasympathetic stimulation to the liver, stomach, and gallbladder is provided by fibers traveling with the
   A) vagus nerve.
   B) greater splanchnic nerve.
   C) less splanchnic nerve.
   D) white rami communicantes.
   E) gray rami communicantes.

   Answer: A

25) A major organ that receives sympathetic stimulation, but not parasympathetic stimulation is the
   A) heart.
   B) liver.
   C) stomach.
   D) lung.
   E) kidney.

   Answer: E

26) The pelvic splanchnic nerves supply
   A) parasympathetic stimulation to smooth muscle in the urinary bladder and reproductive organs.
   B) sympathetic stimulation to smooth muscle in the urinary bladder and reproductive organs.
   C) sympathetic preganglionic fibers to the renal plexus.
   D) sympathetic stimulation to the kidney.
   E) parasympathetic stimulation to the liver and pancreas.

   Answer: A

27) Acetylcholine exerts its effects on postsynaptic cells when it
   A) is broken down by enzymes in the synaptic cleft and the end-products diffuse into the postsynaptic cell.
   B) binds to specific receptors on the postsynaptic cell membrane and changes the permeability of the membrane to particular ions.
   C) diffuses into the postsynaptic cell and changes the pH of the intracellular fluid.
   D) actively transports ions from the synaptic cleft into the postsynaptic cell.
   E) blocks specific receptors to which other neurotransmitters could attach.

   Answer: B

28) Norepinephrine and epinephrine enter the bloodstream when sympathetic stimulation is provided to the

A) adrenal medulla.

B) liver.

C) brain.

D) heart.

E) kidneys.

Answer: A

29) Because the principal active ingredient in tobacco is nicotine, you might expect smoking to enhance the effects of

A) acetylcholine of parasympathetic visceral effectors.

B) acetylcholine on any postganglionic neurons.

C) norepinephrine on the heart and blood vessels.

D) norepinephrine on the limbic system.

E) norepinephrine on most sympathetic visceral effectors.

Answer: B

30) Cholinergic sympathetic postganglionic neurons stimulate the

A) adrenal medulla.

B) basal ganglia.

C) sweat glands.

D) heart.

E) walls of the gastrointestinal tract.

Answer: C

31) Loss of control over bladder and bowel functions in situations involving so-called paradoxical fear is due to

A) the fight-or-flight response.

B) failure of the sympathetic nervous system to respond.

C) failure of the parasympathetic nervous system to respond.

D) inability to produce adequate amounts of acetylcholine to maintain muscle tone.

E) massive activation on the parasympathetic nervous system.

Answer: E

32) You are just about to perform a clinical procedure for the first time and your palms begin to sweat. This is due to
   A) increased sympathetic stimulation of sweat glands possessing alpha receptors.
   B) increased sympathetic stimulation of sweat glands possessing beta receptors.
   C) increased parasympathetic stimulation of sweat glands possessing nicotinic receptors.
   D) increased parasympathetic stimulation of sweat glands possessing muscarinic receptors.
   E) increased sympathetic stimulation of sweat glands possessing muscarinic receptors.

   Answer: A

33) Increased sympathetic stimulation increases the secretion of ALL of the following EXCEPT
   A) glucagon.
   B) renin.
   C) epinephrine.
   D) insulin.
   E) norepinephrine.

   Answer: D

34) Fibers of the ANS traveling with cranial nerve X will stimulate which of the following responses?
   A) increased activity from salivary glands
   B) increased heart rate
   C) increased activity of sweat glands on the palms and soles
   D) decreased force of contraction of cardiac muscle
   E) dilation of blood vessels in the skeletal muscle of the torso

   Answer: D

35) Which of the following is stimulated by the parasympathetic nervous system?
   A) dilation of the bronchioles
   B) erection of the penis
   C) increased gluconeogenesis
   D) dilation of the pupil
   E) increased secretion by sweat glands in the palms and soles

   Answer: B

36) The highest center of autonomic nervous system coordination is the
   A) thalamus.
   B) pons.
   C) hypothalamus.
   D) cerebrum.

   Answer: C

37) What forms the lump or swelling (along the sympathetic trunk) called the paravertebral ganglion?
   A) Receptor cells
   B) Nerve cell bodies of postganglionic sympathetic neurons
   C) Nerve cell bodies of preganglionic sympathetic neurons
   D) Nerve cells bodies of sensory neurons
   E) All answers are correct

   Answer: B

38) Select the neurons that may be adrenergic.
   A) sympathetic postganglionic neurons
   B) preganglionic parasympathetic neurons
   C) parasympathetic postganglionic neurons
   D) motor neurons of the central nervous system
   E) All answers are correct.

   Answer: A

39) Cholinergic fibers include
   A) all sympathetic preganglionic axons.
   B) all parasympathetic preganglionic axons.
   C) all sympathetic postganglionic axons.
   D) all parasympathetic postganglionic axons.
   E) All answers are correct.

   Answer: E

40) Norepinephrine is inactivated in adrenergic synapses by
   A) serotonin.
   B) monoamine oxidase.
   C) acetylcholine esterase.
   D) epinephrine.
   E) neostigmine.

   Answer: B

41) Nicotinic receptors are
   A) a type of acetylcholine receptor.
   B) found on about half of the effectors innervated by the parasympathetic postganglionic neurons.
   C) not found in the sympathetic division of the autonomic nervous system.
   D) a type of adrenergic receptor.
   E) All answers are correct.

   Answer: A

42) Muscarinic receptors are
   A) found on all effectors innervated by the parasympathetic postganglionic axons.
   B) a type of adrenergic receptor.
   C) subdivided into alpha and beta types.
   D) found within sympathetic ganglion.
   E) All answers are correct.

   Answer: A

43) Acetylcholine is inactivated by the enzyme
   A) monoamine oxidase.
   B) catechol-o-methyl transferase.
   C) neostigmine.
   D) esterase.
   E) All answers are correct.

   Answer: A

44) Alpha and beta receptors are associated with
   A) sympathetic neurons.
   B) cholinergic neurons.
   C) muscarinic receptors.
   D) nicotinic receptors.
   E) All answers are correct.

   Answer: A

45) Fight-or-flight is associated with
   A) cholinergic neurons.
   B) sympathetic neurons.
   C) muscarinic neurons.
   D) corticospinal neurons.
   E) All answers are correct.

   Answer: B

46) Rest and digestive activities are associated with
   A) cholinergic neurons.
   B) muscarinic neurons.
   C) sympathetic neurons.
   D) corticospinal neurons.
   E) A and B are correct.

   Answer: A

47) The major control center of the autonomic nervous system receives sensory input from
   A) visceral functions.
   B) olfaction.
   C) gustation.
   D) osmolarity.
   E) All answers are correct.

   Answer: E

48) Autonomic nervous system motor output continues to the spinal cord through relays in the
   A) basal ganglia.
   B) precentral cortex.
   C) postcentral cortex.
   D) reticular formation.
   E) All answers are correct.

   Answer: D

49) Autonomic reflexes regulate controlled conditions in the body such as
   A) blood pressure.
   B) digestion.
   C) defecation.
   D) urination.
   E) All answers are correct.

   Answer: E

50) Autonomic dysreflexia is characterized by
   A) pounding headache.
   B) hyptertension.
   C) flushed, warm skin with profuse sweating about injury level.
   D) anxiety.
   E) All answers are correct.

   Answer: E

MATCHING. Choose the item in column 2 that best matches each item in column 1.

51) Postganglionic fibers serving the heart.

52) Preganglionic fibers from cranial nerve VII.

53) Receives information from greater splanchnic nerve.

54) Receives the lower splanchnic nerve.

A. Celiac ganglion

B. Interior cervical ganglion

C. Submandibular ganglion

D. Inferior mesenteric ganglion

E. Otic ganglion

Answers: 51) B. 52) C. 53) A. 54) D.

55) Muscarinic receptors.

56) Nicotinic receptors.

57) Adrenergic receptors.

58) Beta receptors.

59) Monoamine oxidase effects.

60) Cholinergic receptors.

A. Sympathetic neurons

B. Parasympathetic neurons

C. Both neurons use this

Answers: 55) B. 56) C. 57) A. 58) A. 59) A. 60) C.

61) Associated with the optic nerve.

62) Associated with sublingual salivary glands.

63) Associated with glossophyngeal nerve.

64) Associated with the facial nerve.

A. Submandibular ganglion

B. Pterygopalatine ganglion

C. Otic ganglia

D. Ciliary ganglia

E. All answers are correct

Answers: 61) D. 62) A. 63) C. 64) B.

65) Pupils of the eyes dilate.

66) Fight/flight.

67) Glycogenolysis.

68) Slows the heart.

69) Contraction of the diaphragm.

70) Increases digestion enzymes.

A. Sympathetic responses

B. Parasympathetic responses

C. Associated with both

D. Associated with neither

Answer: 65) A. 66) A. 67) A. 68) B. 69) D. 70) B.

SHORT ANSWER. Write the word or phrase that best completes each statement or answers the question.

71) Effector tissues for autonomic motor neurons are _____, _____, and _____.

Answer: cardiac muscle; smooth muscle; glands

72) An autonomic motor neuron that extends from an autonomic ganglion to a visceral effector is called a(n) _____ neuron.

Answer: postganglionic

73) An autonomic motor neuron that extends from the CNS to an autonomic ganglion is called a(n) _____ neuron.

Answer: preganglionic

74) Stimulation of the posterolateral regions of the hypothalamus would result in a(n) _____ in blood pressure because this region controls the _____ division of the ANS.

Answer: increase; sympathetic

75) Stimulation of the anteromedial regions of the hypothalamus would result in a(n) _____ in secretion and motility of the gastrointestinal tract because this region controls the _____ division of the ANS.

Answer: increase; parasympathetic

76) The main integrating centers for autonomic reflexes are the spinal cord, the brain stem, and the _____.

Answer: hypothalamus

77) Secretion of saliva and gastric juice is increased by the _____ division of the ANS.

Answer: parasympathetic

78) Ejaculation of semen is an effect of the _____ division of the ANS.

Answer: sympathetic

79) Autonomic motor nerves innervating the arterioles of the kidney belong to the _____ division of the ANS.

Answer: sympathetic

80) Sympathetic preganglionic fibers that connect the anterior ramus of a spinal nerve with sympathetic trunk ganglia are collectively called the _____.

Answer: white rami communicantes

81) The effector for sympathetic postganglionic fibers leaving the middle and inferior cervical ganglia is the _____.

Answer: heart

82) Secretion of insulin from the pancreas is increased by stimulation from the _____ division of the ANS.

Answer: parasympathetic

83) Alpha and beta receptors are the receptors for the neurotransmitter _____.

Answer: epinephrine (and norepinephrine)

84) Atropine is a drug that blocks _____ receptors.

Answer: muscarinic

85) Monoamine oxidase inactivates _____.

Answer: epinephrine (and norepinephrine)

86) Parasympathetic cranial outflow has five components: four pairs of ganglia and the plexuses associated with the _____ nerve.

Answer: vagus

87) Activation of the sympathetic division of the ANS also results in the release of hormones from the modified postganglionic cells of the _____.

Answer: adrenal medulla

88) Cholinergic neurons release the neurotransmitter _____.

Answer: acetylcholine

89) _____ acetylcholine receptors are present on the sarcolemma of skeletal muscle fibers.

Answer: Nicotinic

90) Adrenergic neurons release the neurotransmitters _____ and _____.

Answer: epinephrine; norepinephrine

91) _____ receptors for acetylcholine are present on both sympathetic and parasympathetic postganglionic neurons.

Answer: Nicotinic

92) _____ receptors for acetylcholine are present on all effectors innervated by parasympathetic postganglionic neurons.

Answer: Muscarinic

93) Binding of epinephrine to beta receptors on cardiac muscle fibers causes _____.

Answer: increase in rate and force of contraction

94) Binding of epinephrine to beta receptors on the juxtaglomerular cells of the kidneys results in secretion of _____.

Answer: renin

95) Binding of epinephrine to beta receptors on hepatocytes results in _____.

Answer: (increased) glycogenolysis

ESSAY. Write your answer in the space provided or on a separate sheet of paper.

96) Explain why the sympathetic division of the ANS has more widespread and longer-lasting effects than the parasympathetic division.

Answer: A single sympathetic preganglionic neuron synapses with 20 or more postganglionic neurons vs. about five for parasympathetic. The sympathetic neurotransmitters are broken down more slowly than acetylcholine, so postganglionic cells are stimulated longer. The sympathetic division also stimulates release of catecholamines from the adrenal medulla, thus enhancing the sympathetic effects via the endocrine system. Many more visceral effectors have receptors for catecholamines than for acetylcholine.

97) Describe the possible ways in which sympathetic preganglionic neurons may connect with postganglionic neurons.

Answer: The axon may 1) synapse with postganglionic neurons in the first ganglion it reaches, 2) ascend or descend to higher or lower ganglia via sympathetic chains, or 3) continue without synapsing through sympathetic trunk ganglia to prevertebral ganglia via splanchnic nerves.

98) An autonomic neuron releases the neurotransmitter acetylcholine. What can you tell about this neuron's role in the ANS? What possible characteristics can be determined about the postsynaptic cell? Explain your answers.

Answer: The neuron could be a preganglionic neuron in either division of the ANS, a parasympathetic postganglionic neuron, or one of a few sympathetic postganglionic neurons. The postsynaptic cell must possess either nicotinic or muscarinic receptors in order to respond to acetylcholine. This effector cell could be either a postganglionic neuron of either division, a parasympathetic visceral effector, or one of the few sympathetic effectors stimulated by cholinergic neurons.

99) Explain how the ANS regulates blood flow during times of fight-or-flight vs. times of rest/repose.

Answer: The sympathetic ANS serves arterioles and veins in all areas. Increased vasoconstriction or vasodilation occurs with greater sympathetic stimulation depending on the type of receptor present on the smooth muscle cells of the vessel walls. In general, vasodilation occurs in those vessels serving the heart and those skeletal muscles crucial to fight or flight. Vasoconstriction occurs in those vessels serving areas less vital to fighting/fleeing, for example, in skin, digestive organs, and the urinary system. Most vessels do not have parasympathetic innervation.

100) What are splanchnic nerves?

Answer: Sympathetic splanchnic nerves are those that pass through sympathetic trunk ganglia of the abdominopelvic region to extend to and terminate in prevertebral ganglia. The parasympathetic division includes pelvic splanchnic nerves, which are the collective preganglionic outflow from the sacral region.

CHAPTER 18   The Endocrine System

MULTIPLE CHOICE. Choose the one alternative that best completes the statement or answers the question.

1)   Paracrines are
    A) local hormones that act on neighboring cells.
    B) local hormones that act on the cells that produced them.
    C) circulating hormones that are never broken down.
    D) the receptors for steroid hormones.
    E) inactive forms of circulating hormones.

    Answer: A

2)   A chemical grouping of hormones derived from arachidonic acid is the
    A) eicosanoids.
    B) biogenic amines.
    C) proteins.
    D) peptides.
    E) steroids.

    Answer: A

3)   Which of the following hormones works by direct gene activation?
    A) ADH.
    B) hGH.
    C) insulin.
    D) cortisol.
    E) glucagon.

    Answer: D

4)   When a steroid hormone binds to its target cell receptor, it
    A) causes the formation of cyclic AMP.
    B) is converted into cholesterol, which acts as a second messenger.
    C) causes the formation of releasing hormones.
    D) turns specific genes of the nuclear DNA on or off.
    E) alters the membrane's permeability to G proteins.

    Answer: D

5) The compound that most often acts as a second messenger is

   A) cholesterol.

   B) phosphodiesterase.

   C) cyclic AMP.

   D) COMT.

   E) CRH.

   Answer: C

6) The specific effect of a water-soluble hormone on a target cell depends on the

   A) particular gene that is activated.

   B) type of adenylate cyclase present.

   C) source of the phosphate for the phosphorylation reaction.

   D) extent to which the hormone can diffuse into the target cell.

   E) specific protein kinase activated.

   Answer E

7) When a hormone that uses a second messenger binds to a target cell, the next thing that happens is that

   A) phosphodiesterase is activated.

   B) a protein kinase is activated.

   C) a gene is activated in the nucleus.

   D) adenylate cyclase is activated by a G protein.

   E) voltage-regulated ion channels open in the plasma membrane.

   Answer: D

8) Steroids include the following EXCEPT

   A) cortisone.

   B) aldosterone.

   C) estrogen.

   D) insulin.

   E) testosterone.

   Answer: D

9) Amine hormones include the following:

   A) epinephrine.

   B) thyroxin.

   C) dopamine.

   D) All of the answers are correct.

   Answer: D

10) Catecholamines include the following:
   A) epinephrine.
   B) norepinephrine.
   C) adrenalin.
   D) dopamine.
   E) All of the answers are correct.

   Answer: E

11) Phosphodiesterase enzyme inactivates the following:
   A) ATP
   B) ADP
   C) cAMP
   D) GDP
   E) All answers are correct.

   Answer: C

12) Insulinlike growth factors are necessary for the full effect of
   A) insulin.
   B) hGH.
   C) somatostatin.
   D) triiodothyronine.
   E) glucagon.

   Answer: B

13) GnRH directly stimulates the release of
   A) FSH.
   B) estrogen.
   C) testosterone.
   D) DHEA.
   E) All of these are correct.

   Answer: A

14) A hormone that influences an endocrine gland other than its source is called a(n)
   A) autocrine.
   B) paracrine.
   C) eicosanoid.
   D) tropin.
   E) mitogen.

   Answer: D

15) Hypoglycemia is a stimulus for release of
   A) ACTH.
   B) GHRH.
   C) Insulin.
   D) Both A and B.
   E) A, B, and C are correct.

   Answer: D

16) The hypophyseal portal system drains
   A) anterior pituitary.
   B) posterior pituitary.
   C) hypothalamus.
   D) thalamus.
   E) All answers are correct.

   Answer: C

17) Suckling is an important stimulus for release of
   A) oxytocin.
   B) DHEA.
   C) estrogen.
   D) FSH.
   E) LH.

   Answer: A

18) Hormones from the posterior pituitary are released in response to
   A) releasing hormones from the hypothalamus.
   B) nerve impulses from the hypothalamus.
   C) permissive hormones from the pineal gland.
   D) renin from the kidneys.
   E) releasing hormones from the anterior pituitary.

   Answer: B

19) Osmoreceptors in the hypothalamus stimulate secretion of
   A) renin from the kidney in response to low osmotic pressure.
   B) ADH from the hypothalamus in response to low osmotic pressure.
   C) ADH from the hypothalamus in response to high osmotic pressure.
   D) aldosterone from the kidney in response to low osmotic pressure.
   E) aldosterone from the adrenal cortex in response to low osmotic pressure.

   Answer: C

20) Increasing the uptake of iodide by the thyroid gland and increasing the growth of the thyroid gland are two functions of

A) TRH.

B) TSH.

C) T_3.

D) thyroglobulin.

E) thyroid binding globulin.

Answer: B

21) The main target for ADH is the

A) kidney.

B) hypothalamus.

C) uterus.

D) adrenal cortex.

E) posterior pituitary.

Answer: A

22) The primary effect of T_3 and T_4 is to

A) decrease blood glucose.

B) promote the release of calcitonin.

C) promote heat-generating reactions.

D) stimulate the uptake of iodide by the thyroid gland.

E) promote excretion of sodium ions in urine.

Answer: C

23) Increasing synthesis of the enzymes that run the active transport pumpNa^+/K^+ ATPase is the major effect of

A) androgens.

B) glucocorticoids.

C) mineralocorticoids.

D) thyroid hormones.

E) insulin.

Answer: D

24) The primary effect of calcitonin is to

A) increase blood glucose.

B) decrease blood glucose.

C) increase excretion of calcium ions in urine.

D) increase blood calcium.

E) decrease blood calcium.

Answer: E

25) Hormones secreted from the posterior pituitary gland are synthesized by the
    A) anterior pituitary gland.
    B) thyroid gland.
    C) posterior pituitary gland.
    D) hypothalamus.
    E) pineal gland.

    Answer: D

26) Promotion of the formation of calcitriol is a major effect of
    A) parathormone.
    B) aldosterone.
    C) calcitonin.
    D) TSH.
    E) cortisol.

    Answer: A

27) Digestion of the thyroid's colloid uses what cytoplasmic organelle?
    A) golgi apparatus.
    B) endoplasmic reticulum.
    C) lysosome.
    D) nucleus.
    E) All answers are correct.

    Answer: C

28) In the formation of thyroxin, iodine is added to which amino acid
    A) tyrosine.
    B) phenylalanine.
    C) tryptophane.
    D) alanine.
    E) All answers are correct.

    Answer: A

29) Calcitonin is synthesized by
    A) parathyroid.
    B) anterior pituitary.
    C) posterior pituitary.
    D) hypothalamus.
    E) thyroid.

    Answer: E

30) The thyroid gland is located
   A) under the sternum.
   B) behind and beneath the stomach.
   C) in the sella turcica of the sphenoid bone.
   D) in the neck, anterior to the trachea.
   E) in the roof of the third ventricle of the brain.

   Answer: D

31) A primary effect of mineralocorticoids is to promote
   A) increased urine production.
   B) excretion of potassium ions by the kidney.
   C) excretion of sodium ions by the kidney.
   D) decreased blood glucose.
   E) increased secretion of ACTH.

   Answer: B

32) The innermost layer of the adrenal cortex is the
   A) medulla.
   B) delta cells.
   C) zona glomerulosa.
   D) zona reticularis.
   E) zona fasciculata.

   Answer: D

33) The primary stimulus for release of cortisol and corticosterone is
   A) ACTH.
   B) increased levels of blood glucose.
   C) the renin-angiotensin pathway.
   D) increased levels of sodium ions in the blood.
   E) increased levels of calcium ions in the blood.

   Answer: A

34) An increase in blood glucose and an anti-inflammatory effect are important effects of
   A) epinephrine.
   B) glucagon.
   C) corticosterone.
   D) insulin.
   E) ADH.

   Answer: C

35) The stimulus for release of parathyroid hormone is
   A) PRH.
   B) low levels of calcium ions in the blood.
   C) TSH.
   D) nerve impulses from the hypothalamus.
   E) calcitonin.

   Answer: B

36) Rein-angiotensin-aldosterone controls
   A) Na/K.
   B) calcium.
   C) glucose.
   D) tyrosine.
   E) All answers are correct.

   Answer: A

37) The islets of Langerhans are the endocrine portion of the
   A) adrenal cortex.
   B) adrenal medulla.
   C) anterior pituitary gland.
   D) posterior pituitary gland.
   E) pancreas.

   Answer: E

38) The only hormone that promotes anabolism of glycogen, fats, and proteins is
   A) hGH.
   B) insulin.
   C) epinephrine.
   D) aldosterone.
   E) corticosterone.

   Answer: B

39) The primary target for glucagon is the
   A) liver.
   B) hypothalamus.
   C) adrenal cortex.
   D) pancreas.
   E) kidney.

   Answer: A

40) Sympathetic autonomic stimulation increases
    A) glucagon secretion.
    B) secretion from chromaffin cells.
    C) insulin secretion.
    D) Both B and C are correct.
    E) Both A and B are correct.

    Answer: E

41) Increased heart rate and force of contraction are effects of
    A) ADH.
    B) insulin.
    C) cortisol.
    D) epinephrine.
    E) aldosterone.

    Answer: D

42) Dehydroepiandrosterone is a(n)
    A) glucocorticoid.
    B) mineralocorticoid.
    C) androgen.
    D) aldosterone.
    E) cortisone.

    Answer: C

43) Glucocorticoids have the following effects:
    A) gluconeogenesis.
    B) lipolysis.
    C) anti-inflamatory.
    D) depression of immune system.
    E) All answers are correct.

    Answer: E

44) The primary source of estrogens after menopause is the
    A) ovaries.
    B) uterus.
    C) hypothalamus.
    D) thyroid gland.
    E) zona reticularis of the adrenal cortex.

    Answer: E

45) The interstitial cells of the testes are an important target for
   A) FSH.
   B) LH.
   C) GnRH.
   D) oxytocin.
   E) Both A and B are correct.

   Answer: B

46) The role of somatostatin from the pancreas is to
   A) promote secretion of pancreatic digestive enzymes.
   B) promote formation of calcitriol to facilitate calcium absorption from the gastrointestinal tract.
   C) inhibit secretion of insulin and glucagon.
   D) promote secretion of insulin and glucagon.
   E) inhibit the activity of the adrenal cortex.

   Answer: C

47) An increase in glycogenolysis by the liver is an important effect of
   A) glucagon.
   B) insulin.
   C) PTH.
   D) aldosterone.
   E) Both A and B are correct.

   Answer: A

48) The primary stimulus for the release of insulin is
   A) an elevated level of blood glucose.
   B) a decreased level of blood glucose.
   C) insulin-releasing hormone.
   D) insulinlike growth factors.
   E) pancreatic stimulating hormone.

   Answer: A

49) Protein anabolism is promoted by
   A) insulin.
   B) somatotropin.
   C) triiodothyronine.
   D) cortisol.
   E) All of the above except cortisol.

   Answer: E

50) The pineal gland secretes
   A) glucocorticoids.
   B) eicosanoids.
   C) melatonin.
   D) thymosin.
   E) aldosterone.

   Answer: C

MATCHING. Choose the item in column 2 that best matches each item in column 1.

51) Synthesis of lipids.
52) Breakdown of glucose.
53) Glycogen synthesis.
54) Breakdown of lipids.

A. Gluconeogenesis
B. Lipolysis
C. Glycogenesis
D. Lipogenesis
E. Glycolysis

   Answers: 51) D. 52) E. 53) C. 54) B.

55) Synthesizes mineralocorticoids.
56) Synthesizes androgens.
57) Synthesizes glucocorticoids.
58) Synthesizes epinephrine.

A. Zona glomerulosa
B. Zona fasciculata
C. Zona reticularis
D. Medulla
E. All answers are correct

   Answers: 55) A. 56) C. 57) B. 58) D.

59) Secretes glycogen.
60) Secretes somatostatin.
61) Secretes pancreatic polypeptide.
62) Secretes insulin.

A. Alpha cells
B. Beta cells
C. Delta cells
D. C cells
E. F cells

   Answers: 59) A. 60) C. 61) E. 62) B.

63) High blood glucose causes.
64) High blood calcium.
65) Low blood calcium.
66) High blood potassium.

A. Insulin
B. Glucagon
C. Calcitonin
D. Parathyroid hormone
E. Aldosterone

   Answers: 63) A. 64) C. 65) D. 66) E.

67) Release of epinephrine.
68) Release of glucocorticoids.
69) Death of the person.
70) General Adaptation Syndrome.

A. Resistance phase
B. Exhaustion phase
C. Alarm reaction
D. All answers are correct

Answers: 67) C. 68) A. 69) B. 70) D.

SHORT ANSWER. Write the word or phrase that best completes each statement or answers the question.

71) The signs and symptoms of Cushing's syndrome result from hypersecretion of _____.

Answer: cortisol (glucocorticoids)

72) Harmful stress is called _____, while "productive" stress is called _____.

Answer: distress; eustress

73) Tumor angiogenesis factor is so-called because it stimulates the growth of new _____.

Answer: capillaries

74) Erythropoietin is a hormone produced by the _____ to increase _____.

Answer: kidneys; red blood cell production

75) Secretion of _____ by the pineal gland stimulates sleepiness.

Answer: melatonin

76) The primary androgen is _____.

Answer: testosterone

77) Low blood glucose stimulates release of _____ from the pancreas.

Answer: glucagon

78) Acetylcholine released from parasympathetic neurons causes a(n) _____ in secretion of insulin.

Answer: increase

79) Somatostatin is secreted by both the _____ and the _____.

Answer: hypothalamus; pancreas (delta cells)

80) The specialized sympathetic postganglionic neurons that are the hormone-producing cells of the adrenal medulla are called _____.

Answer: chromaffin cells

81) The major androgen secreted by the adrenal cortex is _____.

Answer: dehydroepiandrosterone (DHEA)

82) The most abundant glucocorticoid is _____.

Answer: cortisol (hydrocortisone)

83) The enzyme _____ is secreted by the juxtaglomerular cells of the kidneys in response to a drop in blood pressure.

Answer: renin

84) Aldosterone targets the kidneys to increase reabsorption of _____ into the blood and secretion of _____ into the urine.

Answer: sodium ions; potassium ions

85) The two main targets for angiotensin II are the _____ and the _____.

Answer: adrenal cortex; smooth muscle in walls of arterioles

86) Anti-inflammatory effects are important effects of _____ produced by the adrenal cortex.

Answer: glucocorticoids

87) The _____ glands are located just superior to each kidney.

Answer: adrenal

88) The major hormonal antagonist to glucagon is _____.

Answer: insulin

89) PTH stimulates reabsorption of calcium ions by the kidneys but excretion of _____.

Answer: phosphates

90) Increased secretion of CRH would be stimulated by low levels of the hormone _____.

Answer: cortisol

91) Increased secretion of CRH directly increases the secretion of _____ by the _____ gland.

Answer: ACTH; anterior pituitary

92) Formation of calcitriol in the kidneys is promoted by the hormone _____.

Answer: PTH

93) If levels of T_3 and T_4 are low, one would expect this to stimulate an increase in secretion of both _____ from the _____ and _____ from the _____ to correct the situation.

Answer: TRH; hypothalamus; TSH; anterior pituitary gland

94) The amino acid that is iodinated during the formation of T_3 and T_4 is _____.

Answer: tyrosine

95) The three target tissues/organs for ADH are the _____, the _____, and the _____.

Answer: kidneys; sweat glands; smooth muscle in blood vessel walls

ESSAY. Write your answer in the space provided or on a separate sheet of paper:

96) Describe the role of G proteins in hormone function.

Answer: G proteins provide the link between hormone receptors on the outer surface of the plasma membrane and adenylate cyclase on the inner surface of the membrane. When a water-soluble hormone binds to its receptor, the binding activates a G protein, which in turn activates adenylate cyclase. Adenylate cyclase catalyzes the formation of the second messenger cAMP.

97) Describe and explain the similarities between starvation and diabetes mellitus.

Answer: A starving person is lacking energy-providing nutrient sources and so must use structural components of the body as energy sources. The diabetic consumes adequate nutrients, but due to the lack of insulin, is unable to move glucose into cells and so cannot use the nutrients. In both cases, energy generation is dependent on non-glucose sources, such as fatty acids and amino acids. Mobilization and metabolism of fats and proteins for energy production purposes lead to ketoacidosis, weight loss, and hunger.

98) Compare and contrast the mechanisms of action of lipid-soluble versus water-soluble hormones.

Answer: Upon reaching their targets, lipid-soluble hormones diffuse through the phospholipid bilayer of the target cell membrane and bind to receptors in the cytosol or nucleus. The activated receptor turns a gene on or off, thus regulating synthesis of a protein. Water-soluble hormones bind to membrane receptors, activating a G protein, which activates adenylate cyclase, which converts ATP to the second messenger cAMP, which activates a protein kinase to regulate enzyme action.

99) What is a goiter? Using the appropriate negative feedback loops in your answer, explain how goiters can develop in both hyposecretion and hypersecretion disorders. In these hyposecretion and hypersecretion disorders, would you expect the levels of other hormones involved in the loops to be high or low. Why?

Answer: A goiter is an enlarged thyroid gland. Hyposecretion goiters are usually due to insufficient iodide in the diet. Resulting low levels of thyroid hormones cause increased TRH and TSH until adequate thyroid activity is restored. Graves' disease causes hyperthyroidism by mimicking TSH. Thyroid enlargement occurs, and production of thyroid hormones increases. TRH and natural TSH remain low due to negative feedback, but false TSH pushes thyroid activity.

100) Describe the role of the hypothalamus in the regulation of the pituitary gland.

Answer: The hypothalamus is the integrating center for much sensory input. It secretes releasing and inhibiting hormones, which diffuse into the hypophyseal portal system to regulate secretion of all hormones from the anterior pituitary gland. It also contains receptors that monitor blood osmotic pressure and neural input from reproductive structures. Integration of this input leads to production of ADH and OT by neurosecretory cells. These hormones are then transported through the hypothalamohypophyseal tract to be secreted by exocytosis from the posterior pituitary in response to nerve impulses.

CHAPTER 19   The Cardiovascular System: The Blood

MULTIPLE CHOICE. Choose the one alternative that best completes the statement or answers the question.

1) The buffy coat of centrifuged blood consists mainly of
   A) the ejected nuclei of red blood cells.
   B) gamma globulins.
   C) ruptured red blood cells whose hemoglobin has sunk to the bottom.
   D) white blood cells and platelets.
   E) serum.

   Answer: D

2) The most abundant of the plasma proteins are the
   A) albumins.
   B) hemoglobins.
   C) gamma globulins.
   D) clotting proteins.
   E) alpha globulins.

   Answer: A

3) The total blood volume in an average adult is about
   A) 8 liters.
   B) one liter.
   C) 3 liters.
   D) 5 liters.
   E) 10 liters.

   Answer: D

4) ALL of the following are important functions of plasma proteins EXCEPT
   A) protection against bacteria and viruses.
   B) maintenance of osmotic pressure.
   C) protection against blood loss.
   D) transportation of steroid hormones.
   E) transportation of oxygen.

   Answer: E

5) The percentage of total blood volume occupied by RBCs is called the
   A) red blood cell count.
   B) hematocrit.
   C) white blood cell count.
   D) hemosiderin.
   E) prostacyclin.

   Answer: B

6) Pluripotent stem cells produce
   A) neutrophils.
   B) basophils.
   C) thrombocytes.
   D) lymphocytes.
   E) All answers are correct.

   Answer: E

7) "5 million per cubic millimeter" is a value falling within the normal adult range for the number of
   A) platelets.
   B) all leukocytes.
   C) erythrocytes.
   D) hemoglobin molecules.
   E) neutrophils.

   Answer: C

8) Growth factors that regulate the differentiation and proliferation of red blood cells are called
   A) erythropoietin.
   B) thrombopoietin.
   C) cytokines.
   D) colony-stimulating factors.
   E) interleukins.

   Answer: A

9) The breakdown product of hemoglobin is
   A) biliverdin.
   B) bilirubin.
   C) urobilin.
   D) stercobilin.
   E) All answers are correct.

   Answer: E

10) Thrombopoietin is a hormone that is
    A) produced by the liver to activate thrombin.
    B) produced by the liver to block the action of thrombin.
    C) produced by the liver to stimulate platelet formation.
    D) produced by the kidney to stimulate platelet formation.
    E) produced by the kidney to activate the intrinsic clotting pathway.

    Answer: C

11) Erythropoietin is synthesized by the
    A) red bone marrow.
    B) yellow bone marrow.
    C) erythrocytes.
    D) spleen.
    E) kidneys.

    Answer: E

12) Formed elements that are biconcave discs about 7–8 μm in diameter are
    A) platelets.
    B) band cells.
    C) blast cells that should not be present in circulation.
    D) small lymphocytes.
    E) erythrocytes.

    Answer: E

13) Red blood cells do not consume any of the oxygen they transport because they
    A) do not have the cellular machinery for aerobic ATP production.
    B) cannot remove oxygen from heme once it is attached.
    C) use carbon dioxide in the electron transport chain instead of oxygen.
    D) do not need to generate anti-ATP.
    E) convert oxygen to globin during transport.

    Answer: A

14) The function of hemoglobin is to
    A) protect the DNA of erythrocytes.
    B) produce red blood cells.
    C) produce antibodies.
    D) carry oxygen.
    E) trigger the cascade of clotting reactions.

    Answer: D

15) When carbon dioxide is carried by red blood cells, it is carried in part by
    A) attaching to the iron ion in heme.
    B) the amino acids in globin.
    C) integrins in the plasma membrane.
    D) nitric oxide.
    E) All of these are correct.

    Answer: B

16) Oxygen is transported by red blood cells by binding to
    A) specific receptors on the plasma membrane.
    B) specific receptors within the nucleus of the red blood cell.
    C) the beta polypeptide chain of the globin portion of hemoglobin.
    D) the polypeptide chain of the heme portion of hemoglobin.
    E) the iron ion in the heme portion of hemoglobin.

    Answer: E

17) The primary organs whose macrophages are responsible for phagocytizing worn-out red blood cells are the
    A) spleen and liver.
    B) spleen and kidneys.
    C) liver and kidneys.
    D) lungs and liver.
    E) lungs and kidneys.

    Answer: A

18) The function of transferrin is to
    A) help white blood cells emigrate from blood vessels.
    B) carry iron ions in the bloodstream.
    C) convert fibrinogen to fibrin.
    D) promote differentiation of blast cells in bone marrow.
    E) store iron in the spleen or liver.

    Answer: B

19) Agranular leukocytes that are phagocytic are the
    A) neutrophils.
    B) monocytes.
    C) lymphocytes.
    D) eosinophils.
    E) All of the above except eosinophils.

    Answer: B

20) The most abundant of the leukocytes are the
   A) lymphocytes.
   B) basophils.
   C) monocytes.
   D) neutrophils.
   E) eosinophils.

   Answer: D

21) The formed elements that are fragments of larger cells called megakaryocytes are
   A) neutrophils.
   B) lymphocytes.
   C) erythrocytes.
   D) thrombocytes.
   E) plasma proteins.

   Answer: D

22) When red blood cells wear out, the iron is saved and the remainder of the hemoglobin is
   A) also saved.
   B) excreted as bile pigments.
   C) rearranged into gamma globulins.
   D) broken down by plasmin.
   E) used as an anticoagulant.

   Answer: B

23) Too few white blood cells is called
   A) leukocytosis.
   B) leukopenia.
   C) leukemia.
   D) All answers are correct.

   Answer: B

24) Inflamed tissues attract phagocytes by a phenomenon called
   A) emigration.
   B) adhesion.
   C) chemotaxis.
   D) lysozymes.
   E) All answers are correct.

   Answer: C

25) Platelets initially stick to the wall of a damaged blood vessel because
   A) exposed collagen fibers make a rough surface to which the platelets are attracted.
   B) histamine causes vasoconstriction so that platelets can't fit through the opening.
   C) fibrin threads act like glue to hold them there.
   D) prothrombinase alters the electrical charge of the vessel wall.
   E) the intracellular fluid released by damaged cells in the blood vessel wall has a higher viscosity than plasma.

   Answer: A

26) The role of integrins on the surfaces of neutrophils is to
   A) bind oxygen.
   B) bind carbon dioxide.
   C) attach to endothelium and facilitate emigration.
   D) poke holes in microbial membranes.
   E) stabilize the neutrophils while they undergo mitosis in the bloodstream.

   Answer: C

27) So-called natural killer cells are a form of
   A) thrombocyte.
   B) lymphocyte.
   C) monocyte.
   D) granular leukocyte.
   E) erythrocyte.

   Answer: B

28) A clot in an unbroken vessel is called
   A) thrombosis.
   B) embolism.
   C) agglutination.
   D) adhesion.
   E) aggregation.

   Answer: A

29) The extrinsic pathway of blood clotting is initiated with
   A) fibrinolysin.
   B) prothrombin.
   C) fibrinogen.
   D) thromboplastins.
   E) All answers are correct.

   Answer: D

30) Heparin works as an anticoagulant by
   A) preventing clumping of platelets.
   B) acting as an antagonist to vitamin K.
   C) working with AT-III to interfere with the action of thrombin.
   D) binding calcium ions.
   E) enhancing the production of tissue plasminogen activator.

   Answer: C

31) Dissolution of a blood clot is called
   A) fibrinolysis.
   B) plasminogen.
   C) prostacyclin.
   D) coumadin.
   E) All answers are correct.

   Answer: A

32) The quantity of prothrombin is correlated with vitamin
   A) A.
   B) B.
   C) D.
   D) E.
   E) K.

   Answer: E

33) A floating blood clot is called a
   A) thrombus.
   B) embolus.
   C) agglutinin.
   D) agglutinogen.
   E) All answers are correct.

   Answer: B

34) A person's ABO blood type is determined by antigens present on the
   A) erythrocytes.
   B) platelets.
   C) leukocytes.
   D) gamma globulins.
   E) blood vessels walls.

   Answer: A

35) Type A blood has an isoantibody called
    A) agglutinin A.
    B) agglutinin B.
    C) agglutinin AB.
    D) agglutinin O.
    E) All answers are correct.

Answer: B

36) The initial stimulus for the vasoconstriction that occurs in hemostasis is
    A) prothrombinase.
    B) mechanical damage to the vessel.
    C) thromboxane A2.
    D) plasmin.
    E) thrombin.

Answer: B

37) Production of enzymes such as histaminase that combat the effects of the mediators of inflammation is an important function of
    A) monocytes.
    B) basophils.
    C) plasma cells.
    D) T lymphocytes.
    E) eosinophils.

Answer: E

38) Which of the following indicates a normal differential count in a healthy adult?
    A) 50% neutrophils, 30% lymphocytes, 15% monocytes, 4% eosinophils, 1% basophils
    B) 65% lymphocytes, 20% neutrophils, 10% monocytes, 4% eosinophils, 1% basophils
    C) 65% neutrophils, 25% lymphocytes, 6% eosinophils, 2% monocytes, 2% basophils
    D) 65% neutrophils, 25% lymphocytes, 6% monocytes, 3% eosinophils, 1% basophils
    E) 50% lymphocytes, 30% neutrophils, 10% eosinophils, 8% monocytes, 2% basophils

Answer: D

39) Type O is considered the theoretical universal
    A) recipient because there are no A or B isoantigens on RBCs.
    B) donor because there are no A or B isoantigens on RBCs.
    C) recipient because there are no anti-A or anti-B antibodies in plasma.
    D) donor because there are no anti-A or anti-B antibodies in plasma.
    E) donor because there are no A or B isoantigens on RBCs, nor are there anti-A or anti-B isoantibodies in plasma.

Answer: B

40) Type AB blood contains
   A) only antibody a.
   B) only antibody b.
   C) neither antibody a nor antibody b.
   D) both antibody a and antibody b.

   Answer: C

41) The conversion of soluble fibrinogen into insoluble fibrin requires its enzymatic conversion directly by a substance known as
   A) thromboplastin.
   B) thrombin.
   C) prothrombin.
   D) factor XII (Hagemann factor).
   E) factor XII or fibrin-stabilizing factor.

   Answer: B

42) The symptoms of hemolytic disease of the newborn occur because
   A) the baby has a faulty gene that makes its hemoglobin unable to bind oxygen.
   B) the baby begins making antibodies to its own A or B isoantigens.
   C) anti-Rh antibodies produced by the mother pass the placenta into the bloodstream of the fetus.
   D) the baby is premature and is unable to produce enough plasma proteins to keep osmotic pressure at the correct levels.
   E) the mother took high doses of aspirin during the last trimester of development.

   Answer: C

43) The father is Rh⁻ and the mother is Rh⁺. They have had three children without adverse problems due to the Rh factor. The mother is pregnant again. In terms of Rh factor, the risk to the fetus now within the uterus
   A) is less than before.
   B) is greater than before.
   C) never was a problem.
   D) is the same and remains relatively moderate.

   Answer: C

44) Which blood type is called the universal donor?
   A) A
   B) B
   C) AB
   D) O

   Answer: D

45) People suffering from disorders that prevent absorption of fat from the intestine may suffer uncontrolled bleeding because

A) the fat-soluble vitamin K cannot be absorbed, so levels of prothrombin and other clotting factors drop.

B) fat droplets in the blood normally help platelets and other formed elements stick together during hemostasis.

C) vitamin B12 cannot be absorbed, so there are inadequate formed elements present to produce a clot.

D) fatty acids are necessary for the complete synthesis of fibrin.

E) fat prevents the complete inactivation of plasmin.

Answer: A

46) The purpose for giving RhoGAM to women who have just delivered a child or who have had a miscarriage or abortion is to

A) stimulate contraction of uterine smooth muscle to expel all uterine contents.

B) trigger the clotting cascade to prevent excessive bleeding.

C) block the stimulation of pain receptors.

D) stimulate erythropoiesis to replace red blood cells lost during delivery.

E) block recognition of any fetal red blood cells by the mother's immune system.

Answer: E

47) MHC antigens are

A) the antigens that determine the ABO blood types.

B) viral antigens whose presence indicates presence of infectious disease.

C) proteins encoded by genes present in nucleated cells that must be matched for successful tissue transplantation.

D) the antigens that are attacked in hemolytic disease of the newborn.

E) the molecules to which emigrating leukocytes attach before leaving the bloodstream.

Answer: C

48) A deficiency of Vitamin B_{12} results in

A) iron deficiency anemia.

B) pernicious anemia.

C) hemolytic anemia.

D) thalassemia.

E) aplastic anemia.

Answer: B

49) Destruction of red bone marrow from a variety of drugs results in

    A) iron deficiency anemia.

    B) pernicious anemia.

    C) hemolytic anemia.

    D) aplastic anemia.

    E) All answers are correct.

Answer: D

50) Slightly bluish/dark-purple skin discoloration is called

    A) anemia.

    B) anoxia.

    C) cyanosis.

    D) septicemia.

    E) All answers are correct.

Answer: C

MATCHING. Choose the item in column 2 that best matches each item in column 1.

51)	Blood cell that contains hemoglobin.	A. Neutrophils
52)	Occur as B cells and T cells.	B. Lymphocytes
53)	Associated with allergic reactions.	C. Erythrocytes
54)	Release serotonin.	D. Eosinophils
		E. Basophils

Answers: 51) C. 52) B. 53) D. 54) E.

55)	Enzyme that digests fibrin.	A. Thromboplastin
56)	Forms the protein threads of a blood clot.	B. Fibrinogen
57)	Initiates the extrinsic pathway of blood coagulation.	C. Thrombin
58)	Serves as the catalyst to form fibrin.	D. Fibrinolysin
		E. Fibrin

Answers: 55) D. 56) E. 57) A. 58) B.

59)	Cells that give rise to all blood cells.	A. Erythrocytes
60)	Mature red blood cell.	B. Reticulocytes
61)	Stimulates red blood cell formation.	C. Erythropoietin
62)	Red blood cell that just lost its nucleus.	D. Hemosiderin
		E. Pluripotent stem cell

Answers: 59) E. 60) A. 61) C. 62) B.

63) When hemoglobin breaks down, the first product is

64) Brown pigment in the feces.

65) Yellow pigment in urine.

66) Causes jaundice.

A. Urobilin

B. Stercobilin

C. Biliverdin

D. Bilirubin

E. Ferritin

Answers: 63) C. 64) B. 65) A. 66) D.

67) Destruction of red bone marrow.

68) Deficiency of Vitamin B_{12}.

69) Inadequate absorption of iron.

70) Due to excessive blood loss.

A. Pernicious anemia

B. Hemorrhagic anemia

C. Aplastic anemia

D. Hemolytic anemia

E. Iron-deficient anemia

Answers: 67) C. 68) A. 69) E. 70) B.

SHORT ANSWER. Write the word or phrase that best completes each statement or answers the question.

71) The normal pH of blood ranges from _____ to _____.

Answer: 7.35; 7.45

72) Blood is about _____ % formed elements and _____ % plasma.

Answer: 45; 55

73) The most abundant of the plasma proteins are the _____.

Answer: albumins

74) Males have a higher hematocrit range than females because testosterone stimulates synthesis of the hormone _____.

Answer: erythropoietin

75) Immunoglobulins are produced by _____ , which develop from B lymphocytes.

Answer: plasma cells

76) After birth, hemopoiesis takes place only in _____.

Answer: red bone marrow

77) Thrombopoietin is produced by the _____ to stimulate formation of platelets.

Answer: liver

78) One microliter of blood contains about _____ red blood cells in the adult male.

Answer: 5.4 million

79) Red blood cells are highly specialized for the function of _____.

Answer: oxygen transport

80) Each heme pigment in hemoglobin contains a(n) _____ that can combine reversibly with one oxygen molecule.

Answer: iron ion

81) When hemoglobin passes into the lungs, it releases carbon dioxide and _____ , a gas produced by the endothelial cells lining blood vessels.

Answer: nitric oxide

82) Vasodilation is triggered by _____ as it is released with oxygen from hemoglobin; this triggers a(n) _____ in blood pressure.

Answer: super nitric oxide; decrease

83) Red blood cells survive about _____ days in circulation.

Answer: 120

84) The iron-storage proteins in muscle fibers, liver cells, and macrophages are _____ and _____ .

Answer: ferritin; hemosiderin

85) The non-iron portion of heme is converted first to the green pigment _____ , then to the yellow-orange pigment _____ , which is transported to the liver for secretion in bile.

Answer: biliverdin; bilirubin

86) The stimulus for release of erythropoietin is _____ .

Answer: hypoxia in the kidney

87) The target cells for erythropoietin are the _____ in the red bone marrow.

Answer: proerythroblasts

88) A few days after donating blood at the local blood drive, a person would probably exhibit an elevated level of immature cells called _____ , demonstrating an increased demand for red blood cells.

Answer: reticulocytes

89) On a standard blood smear, cells that stain pale lilac are the _____ , those that stain red-orange are the _____ , and those that stain blue-purple are the _____ .

Answer: neutrophils; eosinophils; basophils

90) The largest of the leukocytes are the _____.

Answer: monocytes

91) _____ are the unique "cell identity marker" proteins protruding from the plasma membranes of white blood cells but not red blood cells.

Answer: Major histocompatibility antigens

92) The normal number of white blood cells per microliter ranges from _____ to _____ cells.

Answer: 5,000; 10,000

93) Mast cells develop from _____ , and intensify the inflammatory response by releasing the chemicals _____ , _____ , and _____ .

Answer: basophils; histamine; heparin; serotonin

94) The attraction of phagocytes to chemicals released by microbes and inflamed tissues is called _____ .

Answer: chemotaxis

95) The molecule _____ released from dense granules of platelets makes other platelets sticky during platelet aggregation.

Answer: ADP

ESSAY. Write your answer in the space provided or on a separate sheet of paper.

96) Explain the proposed role of hemoglobin in the maintenance of blood pressure.

Answer: Hemoglobin releases carbon dioxide and nitric oxide when passing through the lungs. It then picks up oxygen and super nitric oxide, which are then circulated and released to tissues. Release of super nitric oxide causes vasodilation and so a decrease in blood pressure. Nitric oxide causes vasoconstriction and an increase in blood pressure.

97) List and briefly describe the functions of blood.

Answer:

1) Transportation—carries oxygen, carbon dioxide, nutrients, wastes, hormones, and heat.

2) Regulation—helps maintain pH via buffers, body temperature via properties of water in plasma, and water balance via osmotic pressure created by plasma proteins.

3) Protection—via clotting, antibodies, phagocytosis, and complement.

98) Why does damaged endothelium present an increased risk of blood clotting?

Answer: Blood may come in contact with collagen in the surrounding basal lamina, which activates clotting factor XII, which ultimately leads to the formation of fibrin clots. Platelets are also damaged by contact with damaged endothelium and begin their release reaction.

99) On a differential white blood cell count, Ezra is found to have 85 percent neutrophils and an elevated number of band cells. What is the most likely cause of Ezra's high neutrophil count? What benefits are being provided by all these neutrophils? What is the significance of the band cells?

Answer: High neutrophil count indicates bacterial infection (most likely), and increased band cells indicates a rapid turnover of neutrophils with an increased demand for replacements. Neutrophils are phagocytes that, upon engulfing a pathogen, release antimicrobial substances, such as lysozyme, oxidants, and defensins.

100) Describe the negative feedback loop that controls the rate of erythropoiesis. Under what circumstances would you expect the rate of erythropoiesis to be increased? How would it be possible to tell if the rate of erythropoiesis is elevated?

Answer: Hypoxia in the kidney leads to secretion of erythropoietin, which targets proerythroblasts in red marrow to mature into reticulocytes, which enter circulation to increase oxygen-carrying capacity of blood. The rate should be increased in any form of anemia (reduced oxygen-carrying capacity of blood) or when oxygen levels in the external environment are low (e.g., high altitudes). High levels of reticulocytes in circulation indicate an increase in erythropoiesis.

CHAPTER 20   The Cardiovascular System: The Heart

MULTIPLE CHOICE. Choose the one alternative that best completes the statement or answers the question.

1) The space between the parietal and visceral layers of the pericardium is normally filled with
   A) air.
   B) blood.
   C) adipose tissue.
   D) serous fluid.
   E) serum.

   Answer: D

2) The layer of the heart wall responsible for its pumping action is the
   A) fibrous pericardium.
   B) serous pericardium.
   C) epicardium.
   D) myocardium.
   E) endocardium.

   Answer: D

3) Blood flows from the superior vena cava into the
   A) right atrium.
   B) inferior vena cava.
   C) left atrium.
   D) aorta.
   E) pulmonary trunk.

   Answer: A

4) The fossa ovalis is a prominent depression seen in the
   A) wall of the aorta.
   B) interventricular septum.
   C) coronary sinus.
   D) semilunar valves.
   E) interatrial septum.

   Answer: E

5) The myocardium is made of
   A) smooth muscle.
   B) cardiac muscle.
   C) skeletal muscle.
   D) endothelium.
   E) dense connective tissue.
   Answer: B

6) Blood flows from the pulmonary veins into the
   A) pulmonary arteries.
   B) right atrium.
   C) lungs.
   D) left atrium.
   E) left ventricle.
   Answer: D

7) The bicuspid valve is located between the
   A) right ventricle and the aorta.
   B) right ventricle and the pulmonary trunk.
   C) left atrium and the left ventricle.
   D) right and left atria.
   E) right and left ventricles.
   Answer: C

8) There is a semilunar valve between the
   A) right ventricle and the aorta.
   B) right ventricle and the pulmonary trunk.
   C) left atrium and the left ventricle.
   D) right atrium and the right ventricle.
   E) left ventricle and the pulmonary trunk.
   Answer: B

9) Blood flows into the coronary arteries from the
   A) coronary sinus.
   B) superior vena cava.
   C) descending aorta.
   D) pulmonary trunk.
   E) ascending aorta.
   Answer: E

10) The function of the chordae tendineae is to
   A) pull the walls of the ventricles inward during contraction.
   B) open the semilunar valves.
   C) open the AV valves.
   D) prevent eversion of the AV valves during ventricular systole.
   E) hold the heart in place within the mediastinum.

   Answer: D

11) The atrioventricular valves close when the
   A) SA node fires.
   B) atria contract.
   C) vagus nerve stimulates them.
   D) ventricles relax.
   E) ventricles contract.

   Answer: E

12) The atrioventricular valves open when
   A) the chordae tendineae contract.
   B) they are stimulated by the AV node.
   C) ventricular pressure falls below atrial pressure.
   D) atrial pressure falls below ventricular pressure.
   E) the papillary muscles contract.

   Answer: C

13) The role of the papillary muscles is to
   A) secrete pericardial fluid into the pericardial space.
   B) hold the heart in position within the mediastinum.
   C) tighten the chordae tendineae by contracting during ventricular systole.
   D) transmit the action potential to the AV valves.
   E) There is no known function.

   Answer: C

14) All deoxygenated blood returning from the systemic circulation flows into the
   A) right atrium.
   B) right ventricle.
   C) coronary sinus.
   D) left atrium.
   E) left ventricle.

   Answer: A

15) Which of the following represents the correct sequence of parts through which blood moves in passing from the vena cava to the lungs?

 1) right atrium
 2) tricuspid valve
 3) right ventricle
 4) pulmonary valve
 5) pulmonary artery

 A) 1, 2, 3, 4, 5
 B) 3, 1, 2, 4, 5
 C) 4, 1, 2, 5, 3
 D) 2, 5, 4, 3, 1
 E) None of the above.

Answer: A

16) The cusps of the atrioventricular (AV) valves are anchored to the papillary muscles by the
 A) epicardium.
 B) fossa ovalis.
 C) chordae tendineae.
 D) intercalated discs.
 E) pericardium.

Answer: C

17) The left ventricle wall of the heart is thicker than the right wall in order to
 A) accommodate a greater volume of blood.
 B) expand the thoracic cage during diastole.
 C) pump blood with greater pressure.
 D) pump blood through a smaller valve.
 E) pump more blood since the left side feeds the entire body except the lungs.

Answer: C

18) A blockage in the marginal branch of the coronary circulation would most affect the
 A) right atrium.
 B) right ventricle.
 C) pericardium.
 D) left atrium.
 E) left ventricle.

Answer: B

19) The function of intercalated discs is to
    A) initiate the heart beat.
    B) anchor the heart in place within the mediastinum.
    C) prevent eversion of valves.
    D) provide a mechanism for rapid conduction of action potentials among myofibers.
    E) provide an anchoring point for chordae tendineae.

    Answer: D

20) Cardiac muscle cells have less sarcoplasmic reticulum than skeletal muscle cells. The effect of this is that cardiac muscle cells
    A) do not depolarize as quickly.
    B) can function as a single unit.
    C) generate less ATP.
    D) have a smaller intracellular reserve of calcium ions.
    E) are autorhythmic cells.

    Answer: D

21) Which of the following lists the elements of the heart's conduction system in the correct order?
    A) SA node, AV bundle, bundle branches, AV node, conduction myofibers
    B) AV node, SA node, AV bundle, bundle branches, conduction myofibers
    C) SA node, AV node, AV bundle, bundle branches, conduction myofibers
    D) conduction myofibers, AV bundle, bundle branches, AV node, SA node
    E) SA node, AV bundle, AV node, bundle branches, conduction myofibers

    Answer: C.

22) During the normal cardiac cycle, the atria contract when they are directly stimulated by the
    A) SA node.
    B) AV node.
    C) conduction myofibers.
    D) baroreceptors.
    E) vagus nerve.

    Answer: A

23) The initiation of the heart beat is the responsibility of the
    A) cardiovascular center.
    B) baroreceptors.
    C) vagus nerve.
    D) SA node.
    E) fossa ovalis.

    Answer: D

24) A heart beat is normally initiated when
   A) a nerve impulse arrives from the cardiovascular center in the brain.
   B) a critical volume of blood fills the ventricles.
   C) enough sodium and calcium ions leak into the cells of the SA node to reverse their resting potentials.
   D) enough potassium ions leak out of the cells of the SA node to reverse their resting potentials.
   E) the chordae tendineae recoil after being stretched.

   Answer: C

25) On an ECG, depolarization of the atria is represented by the
   A) P wave.
   B) T wave.
   C) QRS complex.
   D) P-Q interval.
   E) S-T segment.

   Answer: A

26) The SA node is located
   A) near the fossa ovalis in the interatrial septum.
   B) in the left atrial wall near the openings of the right pulmonary veins.
   C) in the right atrial wall near the opening of the superior vena cava.
   D) in the aortic wall near the openings to the coronary arteries.
   E) within the pectinate muscles of the right atrium.

   Answer: C

27) Cardiac muscle fibers remain depolarized longer than skeletal muscle fibers because
   A) voltage-gated Na^+ channels close more quickly to trap Na^+ inside longer.
   B) Ca^{++} enters the cytosol from the extracellular fluid to contribute more positive charges slightly after Na^+ have entered.
   C) voltage-gated K^+ channels open at the same time as Na^+ channels, allowing more positively charged K^+ to enter
   D) it takes longer to reach threshold, and the duration of depolarization is directly proportional to the time it takes to reach threshold.
   E) the intercalated discs are very thick relative to the rest of the sarcolemma, it takes longer for K^+ to exit the cell to cause repolarization.

   Answer: B

28) The force of cardiac muscle contraction is influenced primarily by the
   A) number of calcium ions entering the cells through slow channels.
   B) rate at which sodium ions diffuse into the cells.
   C) number of calcium ions that can be stored in the sarcoplasmic reticulum.
   D) duration of the absolute refractory period.
   E) up-and down-regulation of beta adrenergic receptors on the cells.

   Answer: A

29) Opening of voltage-gated K^+ channels in cardiac myofibers allows for
   A) rapid depolarization.
   B) a long refractory period.
   C) repolarization.
   D) rapid conduction between myofibers.
   E) the maintenance of a plateau phase.

   Answer: C

30) On an ECG, depolarization of the ventricles is represented by the
   A) P wave.
   B) T wave.
   C) QRS complex.
   D) P-Q interval.
   E) S-T segment.

   Answer: C

31) An extended P-Q interval on an ECG usually indicates
   A) excessive K to power of (plus) in the extracellular fluid.
   B) a blockage in the conduction system.
   C) damage to the cardiac myofibers in the ventricles.
   D) enlarged ventricles.
   E) nothing unusual or abnormal.

   Answer: B

32) The time the ventricular contractile fibers are depolarized is represented by
   A) P-Q interval.
   B) Q-S interval.
   C) S-T interval.
   D) T interval.
   E) T-P interval.

   Answer: C

33) An artifical pacemaker replaces the function normally served by
   A) the atrioventricular node.
   B) the atrioventricular bundle.
   C) conduction myofibriles.
   D) the sinoatrial node.
   E) all of the above are correct.

   Answer: A

34) The correct sequence of parts that function to carry cardiac impulses is
   1) atrioventricular node
   2) sinoatrial node
   3) purkinje fibers
   4) bundle of His
   5) right and left bundle branches
   A) 1, 2, 3, 4, 5
   B) 2, 4, 5, 3, 1
   C) 3, 2, 1, 5, 4
   D) 2, 1, 4, 5, 3
   E) None of the above.

   Answer: D

35) With each ventricular systole
   A) blood pressure in the aorta increases.
   B) blood pressure in the aorta decreases.
   C) cardiac output decreases.
   D) ventricle becomes filled with blood.
   E) There is no change in blood pressure.

   Answer: A

36) The second heart sound (dupp) is created by the
   A) closing of the atrioventricular valves.
   B) opening of the atrioventricular valves.
   C) closing of the semilunar valves.
   D) opening of the semilunar valves.
   E) vibration of the chordae tendineae during ventricular systole.

   Answer: C

37) Once ventricular pressure falls below atrial pressure
    A) AV valves open.
    B) the heart is in its relaxation period.
    C) the ventricles begin to fill.
    D) blood flows out of the ventricles.
    E) A, B, and C are all correct.

Answer: E

38) The sound associated with the closure of the aortic semilunar valves is best heard near the
    A) superior right point.
    B) superior left point.
    C) inferior right point.
    D) inferior left point.
    E) midpoint of all the above points.

Answer: A

39) The first heart sound (the lubb of lubb-dupp) is caused by the
    A) closure of the semilunar valves.
    B) contraction of the two atria.
    C) contraction of the right ventricle.
    D) closure of the mitral and tricuspid valves.
    E) All answers are correct

Answer: D

40) If acetylcholine is applied to the heart, but cardiac output is to remain constant, which of the following would have to happen?
    A) Stroke volume must increase.
    B) Venous return must decrease.
    C) Force of contraction must decrease.
    D) Rate of conduction of impulses through the AV bundle must increase.
    E) The oxygen content of blood in the coronary circulation must increase.

Answer: A

41) The Frank-Starling Law of the Heart states that
    A) the heart is dependent upon the autonomic nervous system for a stimulus to contract.
    B) the heart contracts to the fullest extent possible for the conditions, or not at all.
    C) cardiac output equals heart rate times stroke volume.
    D) the absolute refractory period for the heart must be longer that the duration of contraction for efficient heart functioning.
    E) a greater force of contraction can occur if the heart muscle is stretched first.

Answer: E

42) The reason that the resting heart rate is lower than the autorhythmic rate of the SA node is that
A) at rest, the AV node is the pacemaker, but the SA node operates only in fight-or-flight situations.
B) norepinephrine keeps calcium ion channels closed.
C) at rest, potassium ions are prevented from re-entering the intracellular fluid.
D) acetylcholine from parasympathetic neurons slows the SA node's rate of initiation of action potentials.
E) the gap junctions between cardiac muscle cells only open in response to input from skeletal muscle proprioceptors during movement.

Answer: D

43) Increased firing of impulses by the sympathetic nervous system would cause
A) an increase in force of contraction of the heart.
B) a shorter absolute refractory period in the SA node.
C) vasoconstriction in coronary circulation.
D) an increase in cardiac output.
E) All of the above except vasoconstriction in coronary circulation.

Answer: E

44) The term *afterload* refers to
A) end-systolic volume.
B) end-diastolic volume.
C) the pressure that must be overcome before semilunar valves can open.
D) the pressure in blood vessels necessary to cause the semilunar valves to close.
E) the maximum possible cardiac output above resting cardiac output.

Answer: C

45) Increased stimulation of the heart by the cardiac accelerator nerves causes
A) stimulation by acetylcholine of muscarinic receptors on the SA node and cardiac muscle fibers of the ventricles.
B) stimulation by norepinephrine of the SA node and of the beta receptors on the cardiac muscle fibers of the ventricles.
C) stimulation by norepinephrine of the SA node but no effect on the cardiac muscle fibers of the ventricles.
D) stimulation by acetylcholine of nicotinic receptors on the SA node and cardiac muscle fibers of the ventricles.
E) stimulation by norepinephrine of the SA node and of the alpha receptors on the cardiac muscle fibers of the ventricles.

Answer: B

46) Stimulation of the heart by autonomic nerve fibers traveling with the vagus nerve causes

   A) increased heart rate and increased ventricular contractility.

   B) decreased heart rate and decreased ventricular contractility.

   C) increased heart rate and no change in ventricular contractility.

   D) decreased heart rate and no change in ventricular contractility.

   E) decreased heart rate and increased ventricular contractility.

   Answer: D

47) Normal resting cardiac output for an average adult is approximately

   A) 70 ml/min.

   B) one liter/min.

   C) 2 liters/min.

   D) 5 liters/min.

   E) 10 liters/min.

   Answer: D

48) The presence of acetylcholine in the heart

   A) retards the passage of the nervous impulse.

   B) decreases the rate of heartbeat.

   C) relaxes the heart muscle.

   D) has an effect opposite to that of histamine.

   E) is associated with the cardiac inhibitory center of the brain.

   Answer: B

49) The heart rate is monitored and regulated in the

   A) thalamus.

   B) diencephalon.

   C) pons.

   D) medulla.

   E) hypothalamus.

   Answer: D

50) Angina pectoris is

   A) an embryonic structure that develops into the aorta.

   B) pain accompanying myocardial ischemia.

   C) the heart's location within the mediastinum.

   D) part of the cardiac conduction system.

   E) the covering of the heart.

   Answer: B

MATCHING. Choose the item in column 2 that best matches each item in column 1.

51) Receives blood from the right ventricle.   A. Vena cava
52) Receives blood from the left ventricle.   B. Coronary artery
53) Carries deoxygenated blood.   C. Aorta
54) Exits from behind the aortic valve.   D. Pulmonary artery
    E. Pulmonary veins

Answers: 51) D. 52) C. 53) A. 54) B.

55) Located in interatrial septum.   A. SA node
56) Located in right atrium.   B. AV node
57) Located in interventricular septum.   C. Bundle of His
58) Located in the left ventricle.   D. Atrioventricular bundles
    E. Purkinje Fibers

Answers: 55) B. 56) A. 57) D. 58) E.

59) Valve between aorta and left ventricle.   A. Mitral valve
60) Valve between pulmonary artery and right ventricle.   B. Tricuspid valve
61) Valve between left ventricle and left atrium.   C. Aortic valve
62) Valve between right ventricle and right atrium.   D. Pulmonary valve
    E. Venous valve

Answers: 59) C. 60) D. 61) A. 62) B.

63) Primary pacemaker.   A. SA node
64) Attaches to the AV node.   B. AV node
65) Secondary pacemaker.   C. Atrioventricular bundle
66) Carries the impulse directly to ventricular muscle fibers.   D. Bundle of His
    E. Purkinje fibers

Answers: 63) A. 64) D. 65) B. 66) E.

67) Causes an increase in heart rate.   A. Epinephrine
68) Causes a decrease in heart rate.   B. Acetyl choline
69) Neurotransmitter associated with the vagus nerve.   C. Sodium
70) Generates an action potential.   D. Potassium
    E. Calcium

Answers: 67) A. 68) B. 69) B. 70) C.

SHORT ANSWER. Write the word or phrase that best completes each statement or answers the question.

71) The membrane surrounding and protecting the heart is the _____.

Answer: pericardium

72) The epicardium is composed of _____ and delicate connective tissue; the myocardium is composed of _____; the endocardium is composed of _____ over a thin layer of connective tissue.

Answer: mesothelium; cardiac muscle; endothelium

73) Cardiac tamponade results from compression created by the buildup of fluid in the _____.

Answer: pericardial space

74) Pectinate muscles are present in the walls of the _____ , and papillary muscles are present in the walls of the _____.

Answer: right atrium; ventricles

75) Valves are composed of _____ covered by _____.

Answer: dense connective tissue; endocardium

76) Most of the base of the heart is formed by the _____ , while the apex of the heart is formed by the _____.

Answer: left atrium; left ventricle

77) The pulmonary semilunar valve is located between the _____ and the pulmonary trunk, and the aortic semilunar valve is located between the _____ and the aorta.

Answer: right ventricle; left ventricle

78) The _____ ventricle has a smaller workload than the _____ ventricle because it only needs to pump blood to the _____.

Answer: right; left; lungs (pulmonary circuit)

79) As pressure in the ventricles rises, the _____ valves open, while the _____ valves close.

Answer: semilunar; AV

80) Blood flows from the superior and inferior venae cavae into the _____.

Answer: right atrium

81) The _____ branch of the left coronary artery distributes oxygenated blood to the walls of the left atrium and left ventricle.

Answer: circumflex

82) Thickenings of the sarcolemmas called _____ hold cardiac muscle fibers together.

Answer: intercalated discs

83) Normally the atria contract in response to an action potential generated by the _____.

Answer: SA node

84) The electrical connection between the atria and ventricles is the _____.

Answer: AV bundle

85) An action potential originating in the SA node ultimately reaches the cardiac muscle fibers of the ventricles via large-diameter cells known as _____.

Answer: conduction myofibers

86) When a site other than the SA node becomes the pacemaker, that site is called a(n) _____ pacemaker.

Answer: ectopic

87) In the heart's conduction system, the action potential is conducted most slowly between the _____ and the _____ , which gives the ventricles a chance to fill as completely as possible.

Answer: AV node; AV bundle

88) The depolarization phase of a cardiac action potential occurs with the opening of voltage-gated fast _____ channels; the plateau phase involves the opening of voltage-gated slow _____ channels; repolarization results when voltage-gated _____ channels open.

Answer: sodium ion; calcium ion; potassium ion

89) Calcium ions allow contraction of cardiac muscle fibers by binding to the regulator protein _____.

Answer: troponin

90) An ECG can be used to determine three conditions: 1) _____; 2) _____; 3) _____.

Answer: status of the conduction pathway; enlargement of the heart; presence of myocardial damage

91) The three upward deflections on an ECG are the _____ , the _____ , and the _____.
The first represents _____ , the second represents _____ , and the third represents _____.

Answer: P wave; QRS complex; T wave; atrial depolarization; ventricular depolarization; ventricular repolarization. (Order doesn't matter, but answers from the first sentence must be matched appropriately with answers in the second sentence.)

92) On an ECG, conduction time from the beginning of atrial excitation to the beginning of ventricular excitation is represented by the _____.

Answer: P-Q interval

93) During the isovolumetric relaxation phase of the cardiac cycle, the AV valves are _____ and the semilunar valves are _____.

Answer: closed; closed

94) The relaxation period of the cardiac cycle includes the three periods known as _____ , _____ , and _____.

Answer: isovolumetric relaxation; rapid ventricular filling; diastasis

95) During the period of ventricular filling, the AV valves are _____ and the semilunar valves are _____.

Answer: open; closed

ESSAY. Write your answer in the space provided or on a separate sheet of paper.

96) A person who is hemorrhaging has a very rapid heart rate. Using your knowledge of the formulas governing cardiac function, explain why. Do you think a rapid heart rate will improve or exacerbate the situation? Why or why not?

Answer: The decrease in blood volume means a decrease in the blood available for end-diastolic volume. If EDV decreases, then stroke volume decreases. If stroke volume decreases, then cardiac output decreases. The increase in heart rate is an attempt to maintain cardiac output (CO = HR × SV), but if the heart rate increases too much, the EDV decreases even further, since ventricular filling time is too short.

97) Explain the importance of calcium ions to cardiac physiology.

Answer: Calcium ions are responsible for the plateau phase following depolarization. As calcium ions enter the cytosol from the ECF through voltage-gated slow channels and from the sarcoplasmic reticulum, depolarization is maintained for 250 msec, thus allowing complete ventricular filling. Calcium also binds to troponin to allow sliding of actin and myosin filaments during contraction, and changes in calcium ion levels alter the force of contraction, and thus, stroke volume.

98) Explain how regular exercise reduces the risk of heart disease.

Answer: Maximum cardiac output and maximum oxygen delivery are increased by increasing the stroke volume while decreasing the heart rate, thus increasing pumping efficiency. Accompanying weight loss reduces blood pressure. Also, HDLs are increased and triglycerides decreased, both helping to decrease the risk of plaque buildup. Increased fibrinolytic activity reduces the risk of intravascular clotting.

99) Describe how the histological structure of the heart supports its function as a pump.

Answer: Cardiac muscle fibers form two separate networks (that of the atria and that of the ventricles). Cells within each network are branched and interconnected via intercalated discs. These discs contain desmosomes and gap junctions. The latter allow action potentials to spread among fibers so that all cells within the network contract as a single functional unit.

100) State the Frank-Starling Law of the Heart, and explain its relationship to cardiac output using the appropriate formula for determining cardiac output.

Answer: A greater preload (EDV, stretch) on cardiac muscle fibers just before they contract increases the force of contraction. Because preload directly affects stroke volume, as EDV increases so does stroke volume; thus force of contraction remains appropriate for the volume in the ventricles. Because CO = HR × SV, as stroke volume increases, so does cardiac output (assuming heart rate stays constant).

CHAPTER 21   The Cardiovascular System: Blood Vessel and Hemodynamics

MULTIPLE CHOICE. Choose the one alternative that best completes the statement or answers the questions.

1) The tunica interna of a blood vessel is made of
   A) smooth muscle.
   B) cardiac muscle.
   C) skeletal muscle.
   D) endothelium.
   E) dense connective tissue.

   Answer: D

2) The layer of a blood vessel wall that determines the diameter of the lumen is the
   A) adventitia.
   B) tunica externa.
   C) tunica interna.
   D) tunica media.
   E) vasa vasorum.

   Answer: D

3) Valves are present in
   A) arteries.
   B) arterioles.
   C) veins.
   D) capillaries.
   E) All of the above except capillaries.

   Answer: C

4) The tunica intima
   A) layer of arterioles is circular smooth muscle fibers.
   B) is in the walls of lymphatic vessels only.
   C) is the inner, endothelial layer of arteries and veins.
   D) is the outer connective tissue layer of bone.
   E) Is none of the above.

   Answer: C

5) Vasoconstriction occurs in the
    A) arteries.
    B) veins.
    C) capillaries.
    D) lymph vessels.
    E) All answers are correct.

    Answer: A

6) Starling's Law of the Capillaries states that
    A) blood flows more slowly through capillaries than arteries or veins because of their smaller diameter.
    B) the volume of fluid reabsorbed at the venous end of a capillary is nearly equal to the volume of fluid filtered out at the arterial end.
    C) if blood pressure is low, blood is diverted around large capillary beds.
    D) if oxygen levels are low, blood is diverted into large capillary beds.
    E) blood pressure in capillaries equals cardiac output divided by resistance.

    Answer: B

7) Net filtration pressure equals
    A) blood hydrostatic pressure plus blood colloid osmotic pressure minus interstitial fluid hydrostatic pressure plus interstitial fluid osmotic pressure.
    B) blood hydrostatic pressure plus interstitial fluid osmotic pressure minus blood colloid osmotic pressure plus interstitial fluid hydrostatic pressure.
    C) blood hydrostatic pressure minus blood colloid osmotic pressure minus interstitial fluid hydrostatic pressure minus interstitial fluid osmotic pressure.
    D) blood hydrostatic pressure plus interstitial fluid hydrostatic pressure minus blood colloid osmotic pressure plus interstitial fluid osmotic pressure.
    E) blood hydrostatic pressure minus interstitial fluid hydrostatic pressure plus blood colloid osmotic pressure minus interstitial fluid osmotic pressure.

    Answer: B

8) Of the pressure involved in determining net filtration pressure, the highest pressure at the arterial end of a capillary is usually
    A) interstitial fluid hydrostatic pressure.
    B) interstitial fluid osmotic pressure.
    C) blood colloid osmotic pressure.
    D) blood hydrostatic pressure.
    E) Blood hydrostatic pressure and blood colloid osmotic pressure are always equally high.

    Answer: D

9) Of the pressure involved in determining net filtration pressure, the highest pressure at the venous end of a capillary is usually

   A) interstitial fluid hydrostatic pressure.

   B) interstitial fluid osmotic pressure.

   C) blood colloid osmotic pressure.

   D) blood hydrostatic pressure.

   E) Blood hydrostatic pressure and blood colloid osmotic pressure are always equally high.

   Answer: C

10) Most fluid and proteins that escape from blood vessels to the interstitial fluid are normally

   A) excreted via the urinary system.

   B) reabsorbed via the urinary system.

   C) returned to the blood via the hepatic portal system.

   D) returned to the blood via the lymphatic system.

   E) absorbed into tissue cells.

   Answer: D

11) If blood hydrostatic pressure equals 30 mm Hg, blood colloid osmotic pressure equals 26 mm Hg, interstitial fluid osmotic pressure equals 5 mm Hg, and interstitial fluid hydrostatic pressure equals 2 mm Hg, the net filtration pressure will be

   A) +7 mm Hg.

   B) −7 mm Hg.

   C) +1 mm Hg.

   D) −1 mm Hg.

   E) +49 mm Hg.

   Answer: A

12) If plasma proteins are lost due to kidney disease, then which of the following pressure changes occur as a direct result?

   A) Blood hydrostatic pressure increases.

   B) Blood colloid osmotic pressure increases.

   C) Interstitial fluid hydrostatic pressure decreases.

   D) Blood colloid osmotic pressure decreases.

   E) Interstitial fluid osmotic pressure decreases.

   Answer: D

13) If lymph channels are blocked, then in areas drained by the blocked vessels
    A) interstitial fluid hydrostatic pressure increases.
    B) blood hydrostatic pressure increases.
    C) blood colloid osmotic pressure increases.
    D) interstitial fluid hydrostatic pressure decreases.
    E) interstitial fluid osmotic pressure decreases.

    Answer: A

14) The diameter of blood vessels most directly affects
    A) venous return.
    B) blood viscosity.
    C) resistance.
    D) heart rate.
    E) stroke volume.

    Answer: C

15) Vascular resistance depends on
    A) size of the lumen of the blood vessel.
    B) viscosity of the blood.
    C) total blood vessel length.
    D) All answers are correct.

    Answer: D

16) An increase in venous return most directly affects
    A) blood pressure.
    B) systemic vascular resistance.
    C) stroke volume.
    D) blood viscosity.
    E) heart rate.

    Answer: C

17) The cardiovascular center is located in the
    A) medulla oblongata.
    B) pons.
    C) thalamus.
    D) cerebellum.
    E) basal ganglia.

    Answer: A

18) Edema results when there is
   A) increased blood hydrostatic pressure.
   B) increased permeability of capillaries.
   C) decreased concentration of plasma proteins.
   D) decreased colloid osmotic pressure.
   E) All answers are correct.

   Answer: E

19) Antidiuretic hormone is often referred to as
   A) angiotensin.
   B) vasopressin.
   C) angiopressin.
   D) atrial natriuretic peptide.
   E) All answers are correct.

   Answer: B

21) The connection between the pulmonary artery and the aorta in fetal circulation is called
   A) ductus venosus.
   B) ductus arteriosus.
   C) ligamentum venosum.
   D) ligamentum arteriosum.
   E) All answers are correct.

   Answer: B

22) The viscosity of blood most directly affects
   A) venous return.
   B) stroke volume.
   C) systemic vascular resistance.
   D) heart rate.
   E) net filtration pressure.

   Answer: C

23) Blood flow increases if
   A) vasodilation increases.
   B) sympathetic stimulation to vessels with alpha adrenergic receptors increases.
   C) blood viscosity increases.
   D) net filtration pressure increases.
   E) parasympathetic stimulation to the heart increases.

   Answer: A

24) The major regulator of regional blood flow in the brain is
   A) autoregulation.
   B) aldosterone.
   C) angiotensin.
   D) the carotid sinus reflex.
   E) the aortic reflex.

   Answer: A

25) The vasomotor region of the cardiovascular center directly controls
   A) heart rate by stimulating the SA node.
   B) stroke volume by regulating total blood volume.
   C) peripheral resistance by changing the diameter of blood vessels.
   D) peripheral resistance by altering blood viscosity.
   E) total blood volume by regulating release of ADH from the posterior pituitary gland.

   Answer: C

26) The function of baroreceptors is to monitor changes in
   A) heart rate.
   B) stroke volume.
   C) peripheral resistance.
   D) blood pressure.
   E) blood viscosity.

   Answer: D

27) Baroreceptors are located in the
   A) wall of the right ventricle.
   B) medulla oblongata.
   C) SA node.
   D) walls of the aorta and carotid arteries.
   E) walls of the capillaries.

   Answer: D

28) Increased levels of epinephrine cause
   A) a decrease in systemic blood pressure due to net vasodilation.
   B) an increase in systemic blood pressure due to net vasoconstriction.
   C) a decrease in systemic blood pressure due to a decrease in force of contraction in cardiac muscle.
   D) an increase in systemic vascular resistance due to an increase in the rate of erythropoiesis.
   E) a decrease in systemic blood pressure due to increased movement of fluid from plasma to the
   · interstitial fluid.

   Answer: B

29) Increased levels of acetylcholine cause

A) a decrease in systemic blood pressure due to net vasodilation.

B) an increase in systemic blood pressure due to net vasoconstriction.

C) a decrease in systemic blood pressure due to a decrease in force of contraction in cardiac muscle.

D) an increase in systemic vascular resistance due to an increase in the rate of erythropoiesis.

E) a decrease in systemic blood pressure due to increased movement of fluid from plasma to the interstitial fluid.

Answer:A

30) A deficiency of ADH would result in

A) reduced venous return.

B) a drop in systemic blood pressure.

C) reduced stroke volume.

D) decreased cardiac output.

E) All of the above are correct.

Answer: E

31) Increased levels of aldosterone cause

A) an increase in venous return because more water is reabsorbed into the blood from the kidneys to increase total blood volume.

B) increased vasodilation because of direct hormone action on the tunica media of arterioles.

C) decreased heart rate because of the decrease in the sodium ion gradient in the SA node.

D) a decrease in systemic blood pressure because of increased fluid loss from the kidneys.

E) an increase in blood viscosity because of the addition of water and sodium ions to the plasma.

Answer: A

32) Hormones that cause vasoconstriction are

A) angiotensin II.

B) angiotensin.

C) epinephrine.

D) vasopressin.

E) All answers are correct.

Answer: E

33) Kristen is allergic to bee venom, and when she goes into anaphylactic shock, epinephrine is administered. This is, in part, because

   A) she has an abnormal parasympathetic response that causes loss of vasomotor tone.

   B) the immune response to the venom clogs the vessels, and epinephrine opens them.

   C) the venom binds to the cardiovascular center neurons and decreases sympathetic activity.

   D) the venom itself alters capillary permeability and increases plasma loss to the interstitial fluid, so epinephrine returns capillary permeability to normal.

   E) histamine, a potent vasodilator, has been released, and epinephrine stimulates vasoconstriction.

   Answer: E

34) "ACE inhibitors" work as antihypertensive drugs by

   A) blocking renin release.

   B) binding to the cardiovascular center to decrease sympathetic stimulation.

   C) blocking release of ADH.

   D) decreasing angiotensin II formation.

   E) binding to aldosterone receptors in the kidney.

   Answer: D

35) The basilar artery is formed by the union of the

   A) internal jugular veins.

   B) vertebral arteries.

   C) internal carotid arteries.

   D) posterior cerebral arteries.

   E) middle cerebral arteries.

   Answer: B

36) The hepatic portal vein is formed by the union of the superior mesenteric vein and the

   A) hepatic artery.

   B) inferior mesenteric vein.

   C) splenic vein.

   D) pancreatic vein.

   E) hepatic vein.

   Answer: C

37) The internal jugular veins receive blood from the

   A) superior vena cava.

   B) brachiocephalic veins.

   C) sigmoid sinuses.

   D) internal iliac veins.

   E) subclavian veins.

   Answer: C

38) The first major branch of the aorta below the diaphragm is the
   A) left subclavian artery.
   B) brachiocephalic trunk.
   C) celiac artery.
   D) superior mesenteric artery.
   E) renal artery.

   Answer: C

39) The primary arteries of the pelvis are the
   A) gonadal arteries.
   B) renal arteries.
   C) femoral arteries.
   D) internal iliac arteries.
   E) external iliac arteries.

   Answer: D

40) The left colic, sigmoid, and superior rectal arteries are all branches of the
   A) aorta.
   B) brachiocephalic trunk.
   C) celiac artery.
   D) superior mesenteric artery.
   E) inferior mesenteric artery.

   Answer: E

41) Blood in the vertebral veins flows next into the
   A) sigmoid sinuses.
   B) brachiocephalic veins.
   C) common carotid veins.
   D) internal jugular veins.
   E) subclavian veins.

   Answer: B

42) Anterior to the elbows, the median cubital veins connect the
   A) axillary and brachial veins.
   B) cephalic and brachial veins.
   C) radial and ulnar veins.
   D) cephalic and basilic veins.
   E) basilic and brachial veins.

   Answer: D

43) If the inferior vena cava or hepatic portal vein becomes obstructed, blood can be returned from the lower body to the superior vena cava via the
   A) azygos vein.
   B) hepatic vein.
   C) superior mesenteric vein.
   D) brachiocephalic vein.
   E) None of these—there is no way blood from the lower body can reach the superior vena cava.

   Answer: A

44) Which of the following flows directly into the inferior vena cava?
   A) internal iliac vein.
   B) hepatic portal vein.
   C) superior vena cava.
   D) hepatic vein.
   E) All of these are correct.

   Answer: D

45) Blood in the great saphenous vein flows into the
   A) femoral vein.
   B) popliteal vein.
   C) peroneal vein.
   D) inferior vena cava.
   E) internal iliac vein.

   Answer: A

46) Deoxygenated blood in the fetal circulation is carried in the
   A) umbilical vein.
   B) umbilical artery.
   C) ductus venosus.
   D) Both A and C are correct.
   E) Both B and C are correct.

   Answer: B

47) A superficial vein in the leg that is frequently used for prolonged administration of IV fluids is the
   A) femoral.
   B) popliteal.
   C) great saphenous.
   D) common iliac.
   E) peroneal.

   Answer: C

48) Blood flows into the left common carotid artery from the
    A) arch of the aorta.
    B) brachiocephalic trunk.
    C) right common carotid artery.
    D) left subclavian artery.
    E) left internal carotid artery.

    Answer: A

49) Vessels that are part of the cerebral arterial circle include the
    A) superior and inferior sagittal sinuses.
    B) external carotid arteries.
    C) internal carotid arteries.
    D) vertebral arteries.
    E) All of the above except the sagittal sinuses.

    Answer: C

50) Blood flows directly into the superior vena cava from the
    A) inferior vena cava.
    B) brachiocephalic veins.
    C) coronary sinus.
    D) internal jugular veins.
    E) axillary veins.

    Answer: B

MATCHING. Choose the item in column 2 that best matches each item in column 1.

51) Endothelium is associated with   A. Artery
52) Elastic lamina is associated with   B. Vein
53) Diastolic pressure is due mainly to   C. Capillary
54) Cross-sectional area is greatest in   D. Lymph vessel
                                               E. All are correct

    Answers: 51) E. 52) A. 53) A. 54) C.

55) Vagus nerve is associated with   A. Sympathetic nervous system
56) Norepinephrine is associated with   B. Parasympathetic nervous system
57) Acetyl choline is the neurotransmitter for   C. All answers are correct
58) Cardioaccelerator nerves are associated with

    Answers: 55) B. 56) A. 57) B. 58) A.

59) Due primarily to blood proteins.   A. Blood hydrostatic pressure

60) Backup of lymphatic drainage will increase   B. Interstitial fluid osmotic pressure

61) Systolic blood pressure relates to   C. Interstitial fluid pressure

62) Increased permeability of capillaries increases   D. Blood colloid osmotic pressure

                                               E. All answers are correct

Answers: 59) D. 60) D. 61) A. 62) D.

63) Renin is associated with   A. Antidiuretic hormone

64) Vasopressin is associated with   B. Epinephrine

65) Adrenal gland medulla secretes   C. Angiotensin

66) Affect hormonal regulation of blood pressure.   D. Atrialnatriuretic peptide

                                               E. All answers are correct

Answers: 63) C. 64) A. 65) B. 66) E.

67) Returns oxygenated blood from placenta.   A. Ductus venosus

68) Opening in the septum of the atria.   B. Ductus arteriosus

69) Becomes the ligamentum venosus after birth.   C. Foramen ovale

70) Connects pulmonary trunk to aorta.   D. Umbilical arteries

                                               E. Umbilical vein

Answers: 67) D. 68) C. 69) A. 70) B.

SHORT ANSWER. Write the word or phrase that best completes each statement or answers the question.

71) The blood vessels that serve the blood vessel walls are called the _____.

Answer: vasa vasorum

72) The layer of a blood vessel wall that is closest to the lumen is the tunica _____, which is composed of _____, a basement membrane, and elastic tissue.

Answer: interna; endothelium

73) The tunica media of a blood vessel wall is composed of _____ fibers and _____ fibers.

Answer: elastic; smooth muscle

74) Increased sympathetic stimulation to blood vessel walls typically causes _____.

Answer: vasoconstriction

75) Two examples of elastic arteries include the _____ and the _____ arteries.

Answer: any two of the following: aorta, brachiocephalic, common carotid, subclavian, vertebral, pulmonary, or common iliac

76) A vessel that emerges from an arteriole to supply a capillary bed is called a(n) _____.

Answer: metarteriole.

77) The ring of smooth muscle fibers that regulates blood flow into a capillary bed is called a(n) _____.

Answer: precapillary sphincter

78) Three types of capillaries include the _____, found in skeletal and smooth muscle, the _____, found in kidneys, and the _____, found in the liver and red bone marrow.

Answer: continuous capillaries; fenestrated capillaries; sinusoids

79) The most important mechanism of capillary exchange is _____.

Answer: simple diffusion

80) Pressure-driven movement of fluid and solutes from the blood capillaries into the interstitial fluid is called _____, while pressure-driven movement from the interstitial fluid into capillaries is called _____.

Answer: filtration; reabsorption

81) The two pressures that promote filtration are _____ and _____, while the main pressure that promotes reabsorption is _____.

Answer: blood hydrostatic pressure; interstitial fluid osmotic pressure; blood colloid osmotic pressure

82) Blood colloid osmotic pressure is a force caused by the _____.

Answer: colloidal suspension of large plasma proteins

83) Blood flows most slowly through the _____.

Answer: capillaries

84) An abnormal increase in formed elements, such as is seen in leukemia, increases resistance because it increases _____.

Answer: blood viscosity

85) The primary factor affecting blood pressure under normal circumstances is _____.

Answer: blood vessel diameter/radius

86) The part of the cardiovascular center that regulates blood vessel diameter is the _____.

Answer: vasomotor center

87) General systemic blood pressure is regulated via the _____ reflex; nerve impulses generated by baroreceptors in this reflex are conducted via the _____ nerve to the cardiovascular center.

Answer: aortic; vagus

88) _____ is a hormone that is a potent vasoconstrictor and that stimulates secretion of aldosterone.

Answer: Angiotensin II

89) In systemic circulation blood vessels _____ in response to a decrease in oxygen concentration, but in the pulmonary circulation, blood vessels _____ in response to a decrease in oxygen concentration.

Answer: dilate; constrict

90) Failure of the cardiovascular system to deliver enough oxygen and nutrients to meet cellular metabolic needs is called _____.

Answer: shock

91) Hemorrhage can lead to _____ shock, while a myocardial infarction leads to _____ shock.

Answer: hypovolemic; cardiogenic

92) Shock stemming from the action of certain bacterial toxins on blood vessels is called _____ shock.

Answer: septic

93) A pulse taken posterior to the knee is taken in the _____ artery.

Answer: popliteal

94) The branches of the aorta that carry blood to the lower limbs are the _____ arteries.

Answer: common iliac

95) Blood in the renal veins flows next into the _____.

Answer: inferior vena cava

ESSAY. Write your answer in the space provided or on a separate sheet of paper.

96) What is the function of capillaries? How are their structure and arrangement especially suited to their function?

Answer: The function of capillaries is to permit exchange of nutrients, wastes, and gases between the blood and cells via the interstitial fluid. Walls of capillaries consist of a single layer of endothelial cells, so exchange is favored by the short distance traveled. Also, capillaries form networks that serve to increase surface area. This large cross-sectional area favors slow blood flow. Both slow flow and large surface area facilitate capillary exchange.

97) What is venoconstriction and what is its significance?

Answer: Venoconstriction is the constriction of veins in response to increased sympathetic stimulation. Systemic veins and venules contain about 60 percent of the total blood volume. When they constrict, blood in these reservoirs is decreased and flow to skeletal muscles is increased. Such a diversion of flow might occur in cases of hemorrhage, for example.

98) Identify and describe the effects of the forces (pressures) affecting movement of fluids and solutes across a capillary membrane.

Answer: Blood hydrostatic pressure is generated by the pumping action of the heart and is the force of water in blood plasma against the inner walls of vessels. This pressure favors filtration. Blood colloid osmotic pressure is generated by the suspension of plasma proteins and is a reabsorptive force. Interstitial fluid osmotic pressure is generated by proteins that leak into the interstitial fluid. This force tends to promote filtration. Interstitial fluid hydrostatic pressure opposes blood colloid osmotic pressure, but is generally considered to be negligible under normal circumstances.

99) Describe the mechanisms by which edema can develop.

Answer: Blood hydrostatic pressure may increase due to high blood volume created by excretion problems or due to increased venous pressure created by heart failure or intravascular clotting. Blood colloid osmotic pressure may be low if plasma protein levels are reduced by malnutrition, burns, or liver/kidney disease. Interstitial fluid osmotic pressure may be high if plasma proteins leak out of vessels and water follows in conditions of inflammation or if lymph flow is blocked, preventing return of proteins and fluid to blood. (Formula for net filtration pressure is helpful in explaining answers.)

100) Identify and discuss the factors that contribute to systemic vascular resistance.

Answer:
1) Blood viscosity ratio of formed elements and proteins to plasma; increasing viscosity via increasing formed elements or decreasing plasma volume increases resistance.
2) Total blood vessel length directly proportional to resistance; increasing length of circuit (by adding new blood vessels to serve added tissue) increases resistance.
3) Diameter/radius of blood vessels as major effect on resistance; increased diameter decreases resistance, thus increasing flow; controlled by ANS; small vessels have greater effect because more surface area is in contact with blood.

CHAPTER 22   The Lymphatic and Immune System and Resistance to Disease

MULTIPLE CHOICE. Choose the one alternative that best completes the statement or answers the question.

1) The composition of lymph is most similar to
   A) plasma.
   B) serum.
   C) cytosol.
   D) interstitial fluid.
   E) intestinal juice.

   Answer: D

2) You would expect anchoring filaments to open spaces between endothelial cells in lymph capillaries when
   A) blood hydrostatic pressure is low.
   B) blood colloid osmotic pressure is high.
   C) interstitial fluid hydrostatic pressure is high.
   D) interstitial fluid osmotic pressure is low.
   E) sympathetic stimulation to the filaments increases.

   Answer: C

3) The thoracic duct empties lymph into the
   A) right lymphatic duct.
   B) cisterna chyli.
   C) left subclavian vein.
   D) ventricles of the brain.
   E) right atrium of the heart.

   Answer: C

4) The cisterna chyli is
   A) the point at which lymph is returned to venous blood.
   B) a dilation at the beginning of the thoracic duct.
   C) the embryonic thymus gland.
   D) the organ that produces the largest amount of lymph.
   E) the array of lacteals associated with the small intestine.

   Answer: B

5)   People who are confined to bed for long period of time often develop edema because
   A) their blood pressure becomes elevated, forcing more fluid into interstitial spaces as blood hydrostatic pressure rises.
   B) lack of motor activity leads to reduced sympathetic stimulation to lymphatic vessels, so lymph tends to pool.
   C) without skeletal muscle contraction to force lymph through lymphatic vessels, fluid tends to accumulate in interstitial spaces.
   D) reduced vasomotor tone allows proteins to leak from plasma, and water follows the osmotic gradient.
   E) heart rate and force of contraction are reduced, so the pressure gradient is insufficient to maintain lymph flow.

   Answer: C

6)   Which of the following correctly lists the structures according to the sequence of fluid flow?
   A) lymphatic capillaries, interstitial spaces, blood capillaries, lymphatic vessels, lymphatic ducts, subclavian veins
   B) blood capillaries, lymphatic vessels, interstitial spaces, lymphatic capillaries, lymphatic ducts, subclavian veins
   C) blood capillaries, interstitial spaces, lymphatic capillaries, lymphatic ducts, lymphatic vessels, subclavian veins
   D) blood capillaries, interstitial spaces, lymphatic capillaries, lymphatic vessels, lymphatic ducts, subclavian veins
   E) blood capillaries, interstitial spaces, lymphatic vessels, lymphatic ducts, lymphatic capillaries, subclavian veins

   Answer: D

7)   Which of the following lists the structures in the correct order of lymph flow through a lymph node?
   A) afferent lymphatic vessels, medullary sinuses, trabecular sinus, subcapsular sinus, efferent lymphatic vessels
   B) efferent lymphatic vessels, trabecular sinus, subcapsular sinus, medullary sinuses, afferent lymphatic vessels
   C) afferent lymphatic vessels, subcapsular sinus, trabecular sinus, medullary sinuses, efferent lymphatic vessels
   D) efferent lymphatic vessels, medullary sinuses, trabecular sinus, subcapsular sinus, afferent lymphatic vessels
   E) Either C or D is correct because lymph can flow either way through a lymph node.

   Answer: C

8) One known function of the reticular epithelial cells of the thymus is to
   A) produce and secrete thymic hormones.
   B) produce and secrete antibodies.
   C) act as antigen-presenting cells.
   D) produce and secrete interleukin-1.
   E) differentiate into natural killer cells.

   Answer: A

9) Specialized lymphatic capillaries called lacteals are found in
   A) the stomach.
   B) leg.
   C) small Intestine.
   D) kidney.
   E) large intestine.

   Answer: C.

10) Which of the following is considered to be a primary lymphatic organ?
    A) red bone marrow
    B) spleen
    C) any lymph node
    D) pharyngeal tonsil
    E) liver

    Answer: A

11) Which person most likely has the largest thymus gland?
    A) third trimester fetus
    B) 2-year-old
    C) 12-year-old
    D) 25-year-old
    E) 65-year-old

    Answer: C

12) T cells and B cells are
    A) phagocytes.
    B) antibodies.
    C) lymphocytes.
    D) complement proteins.
    E) both phagocytes and lymphocytes.

    Answer: C

13) B cells proliferate and differentiate into plasma cells in the
   A) liver.
   B) bloodstream.
   C) germinal centers of lymph nodes.
   D) red pulp of the spleen.
   E) Both C and D are correct.

   Answer: C

14) Which of the following is NOT a function of the spleen?
   A) site of stem cell maturation into T and B cells
   B) destruction of blood-borne pathogens by macrophages
   C) removal of worn-out blood cells and platelets by macrophages
   D) storage of platelets
   E) hemopoiesis during fetal development

   Answer: A

15) The largest lymphatic organ is the
   A) lymph node.
   B) thymus gland.
   C) tonsil.
   D) spleen.
   E) appendix.

   Answer: D

16) A chemical that is produced by virus-infected cells and released to provide nonspecific antiviral protection to neighboring cells is
   A) transferrin.
   B) interleukin-1.
   C) histamine.
   D) interleukin-2.
   E) interferon.

   Answer: E

17) Lysozyme is
   A) an enzyme found in body fluids that flow over epithelial surfaces that destroys certain bacteria.
   B) a type of antibody that makes something more recognizable to a phagocyte.
   C) a cytokine produced by helper T cells.
   D) one of the self-antigens on the surface of antigen-presenting cells.
   E) an antihistamine released by eosinophils.

   Answer: A

18) Which of the following would most likely increase a person's risk of invasion by pathogenic microbes?
   A) increased urine flow
   B) loss of epidermal tissue
   C) increased action of cilia
   D) increased intestinal motility
   E) decreased interstitial fluid osmotic pressure

   Answer: B

19) The function of interferons is to
   A) breakdown bacterial cell walls.
   B) fragment bacterial DNA.
   C) opsonize microbes.
   D) increase capillary permeability.
   E) prevent viral replication.

   Answer: E

20) Which of the following is a nonspecific mechanism of resistance?
   A) activation of the complement system via the alternative pathway
   B) binding of an allergen to IgE molecules on mast cells
   C) a delayed hypersensitivity response to poison ivy
   D) cloning of B cells in response to a measles vaccine
   E) a transfusion reaction between incompatible blood types

   Answer: A

21) The swelling associated with inflammation is caused by
   A) the large numbers of phagocytes attracted to the area.
   B) blockage of blood flow in capillaries by infecting bacteria.
   C) movement of fluid out of capillaries due to increased capillary permeability.
   D) the accumulation of intracellular fluid released by damaged cells.
   E) a larger volume of blood in dilated vessels.

   Answer: C

22) After phagocytosis, which cellular organelle attaches to the phagosome and digests it?
   A) nucleus
   B) Golgi apparatus
   C) lysosome
   D) peroxisome
   E) endoplasmic reticulum

   Answer: C

23) The following substances contribute to vasodilation, increased permeability, and other aspects of inflammation EXCEPT

    A) histamine.

    B) kinins.

    C) prostaglandins.

    D) epinephrine.

    E) leukotrienes.

Answer: D

24) Plasma cells are a form of

    A) helper T cell.

    B) B cell.

    C) killer T cell.

    D) macrophage.

    E) complement.

Answer: B

25) Antibody-mediated immunity is most effective against

    A) fungi.

    B) intracellular viruses.

    C) cancer cells.

    D) antigens in body fluids.

    E) foreign tissue transplants.

Answer: D

26) Antibodies are

    A) plasma cells.

    B) B lymphocytes.

    C) T lymphocytes.

    D) gamma globulin glycoproteins.

    E) cytokines released by macrophages.

Answer: D

27) The significance of haptens in immune responses is that

    A) they must be presented to T cells along with foreign antigens to trigger T cell cloning.

    B) they can combine with larger proteins in the body to become immunogenic, but are reactive without the larger protein.

    C) they determine the immunoglobulin class to which an antibody molecule belongs.

    D) they prevent entry of viruses into cells.

    E) they are the T and B cells that have undergone negative selection.

Answer: B

28) A hapten is
   A) a small substance that has reactivity but lacks immunogenicity.
   B) the heavy chain of an immunoglobulin molecule.
   C) the antigen binding site of an immunoglobulin molecule.
   D) the part of a lymphoid organ where antigens are processed.
   E) a self-antigen genetically encoded by the major histocompatibility complex.

   Answer: A

29) As part of the processing of exogenous antigens, an antigen-presenting cell digests an antigen into fragments and also synthesizes and packages
   A) alpha interferon.
   B) MHC-II antigens.
   C) histamine.
   D) antibodies.
   E) thymic hormones.

   Answer: B

30) The success of an organ or tissue transplant depends on
   A) histocompatibility.
   B) genetic recombination.
   C) haptens.
   D) cytokines.
   E) All answers are correct.

   Answer: A

31) Presentation of an endogenous antigen bound to an MHC-I molecule signals that
   A) everything is normal.
   B) the cell has differentiated into a plasma cell.
   C) a helper T cell is ready to secrete interleukin-2.
   D) a phagocyte is "full" and cannot ingest more antigens.
   E) a cell has been infected.

   Answer: E

32) In cell-mediated immunity, the antigenic cell/molecule is destroyed by
   A) killer T cells.
   B) mast cells.
   C) opsonizing antibodies.
   D) complement.
   E) plasma cells.

   Answer: A

33) During specific immunity, competent T cells are activated by
    A) plasma cells.
    B) complement.
    C) antibodies.
    D) interleukin-1.
    E) histamine.
    Answer: D

34) Cytotoxic T cells recognize antigens combined with
    A) interleukin-1.
    B) interleukin-2.
    C) MHC-I antigens.
    D) MHC-II antigens.
    E) complement proteins.
    Answer: C

35) Antibodies are produced by
    A) macrophages.
    B) killer T cells.
    C) neutrophils.
    D) mast cells.
    E) plasma cells.
    Answer: E

36) The immunoglobulin class of an antibody molecule is determined by the
    A) structure of the L chains.
    B) structure of the variable region.
    C) structure of the constant region of the H chains.
    D) function of the molecule.
    E) type of antigen that stimulates production of the antibodies.
    Answer: C

37) The process of coating an antigenic microbe with antibodies to make it more susceptible to phagocytosis is called
    A) chemotaxis.
    B) opsonization.
    C) cloning.
    D) anergy.
    E) inflammation.
    Answer: B

38) Which of the following is an example of a specific immune response?
   A) release of histamine from damaged cells
   B) adherence of a macrophage to a microbe
   C) release of interferon from virus-infected cells
   D) opsonization of an antigen by IgG molecules
   E) cytolysis of microbes by complement proteins

   Answer: D

39) The most common structural class of antibody molecules is
   A) IgA.
   B) IgM.
   C) IgG.
   D) IgD.
   E) IgE.

   Answer: C

40) Immunoglobulins that circulate in the interstitial spaces and bloodstream attached to mast cells and basophils are classed as
   A) IgA.
   B) IgM.
   C) IgG.
   D) IgD.
   E) IgE.

   Answer: E

41) The immunoglobulin important for providing passively acquired immunity to the fetus in utero is
   A) IgA.
   B) IgM.
   C) IgG.
   D) IgD.
   E) IgE.

   Answer: C

42) ALL of the following are characteristic of the secondary antibody response EXCEPT
   A) proliferation of differentiation of memory cells.
   B) development of a higher antibody titter than in the primary response.
   C) production of antibodies with a higher affinity for the antigen than those in the primary response.
   D) predominant production of IgM antibodies.
   E) response occurring within hours of exposure.

   Answer: D

43) "Teaching" lymphocytes to recognize self from non-self antigens is the function of the
   A) plasma cell.
   B) spleen.
   C) thymus.
   D) macrophage.
   E) liver.

   Answer: C

44) Possibly fatal constriction of the bronchioles and a rapid drop in blood pressure are typical of
   A) anaphylactic hypersensitivity.
   B) phagocytosis when it occurs too rapidly.
   C) overproduction of memory B cells.
   D) delayed hypersensitivity.
   E) immune complex hypersensitivity.

   Answer: A

45) Giving someone an intravenous injection of immunoglobulin would
   A) protect him from a specific disease by giving him passively acquired immunity.
   B) cause him to produce his own antibodies to the pathogen causing the disease.
   C) protect him for several years.
   D) trigger formation of memory B cells that can make antibodies to protect him from this disease in the future.
   E) All of the above are correct.

   Answer: A

46) Which of the following occurs in delayed hypersensitivity?
   A) The complement system is activated by killer T cells.
   B) Memory B cells produce antibodies within hours of a second exposure to an antigen.
   C) Sensitized T cells migrate to the antigen site within 48–72 hours.
   D) IgE antibodies cause histamine to be released from mast cells when they combine with an antigen.
   E) Immune complexes precipitate into joint and kidney tubules.

   Answer: C

47) Receiving an immunization with an altered form of the tetanus toxin results in
   A) naturally acquired active immunity.
   B) naturally acquired passive immunity.
   C) artificially acquired active immunity.
   D) artificially acquired passive immunity.
   E) no response, because altered toxins cannot act as antigens.

   Answer: C

48) The term *immunological tolerance* refers to
   A) inability of the immune system to respond to a particular antigen.
   B) lack of reactivity to peptide fragments from one's own proteins.
   C) the maximum dosage of an antigen to which one can be exposed without initiating an immune response.
   D) the actual number of microbes necessary to cause signs and symptoms of a disease.
   E) the ability of memory cells to recognize an antigen from prior exposure.

   Answer: B

49) The term *anergy* refers to
   A) immobilization of a bacterium by specific antibodies.
   B) making a microbe more susceptible to phagocytosis by coating it with antibodies.
   C) attraction of phagocytes to an area of tissue damage by chemicals released from damaged cells.
   D) the lack of reactivity to peptide fragments from one's own proteins.
   E) failure of a self-reactive lymphocyte to respond to antigenic stimulation.

   Answer: E

50) The antigen-binding site of an antibody molecule is contained in the
   A) hinge region.
   B) disulfide bonds.
   C) constant region of the L chains.
   D) constant region of the H chains.
   E) variable regions of the H and L chains.

   Answer: E

MATCHING. Choose the item in column 2 that best matches each item in column 1.

51) Cell-mediated immunity is associated with
52) Lymph nodes contain
53) Thymus gland produces
54) Helper cells are associated with

A. B Lymphocytes
B. T Lymphocytes
C. Both types are correct

   Answers: 51) B. 52) C. 53) B. 54) B.

55) Bean-shaped structure located along lymphatic vessel.
56) Produces pre-T and B cells.
57) Single largest mass of lymphatic tissue.
58) Responsible for maturation of T cells.

A. Red bone marrow
B. Thymus
C. Lymph nodes
D. Spleen
E. Tonsils

   Answers: 55) C. 56) A. 57) D. 58) B.

59) Found in sweat, tears, saliva.
60) Found in pentamers.
61) Most abundant of all antibodies.
62) Involved in allergic reactions.

A. IgG
B. IgA
C. IgM
D. IgD
E. IgE

   Answers: 59) B. 60) C. 61) A. 62) E.

63) Promotes proliferation of helper T cells.
64) Important for turning off immune responses.
65) Stimulates accumulation of neutrophils and macrophages.
66) Cytokine.

A. Interleukin-1
B. Tumor necrosis factor
C. Transforming growth factor
D. Gamma-interferon
E. All answers are correct.

   Answers: 63) A. 64) C. 65) B. 66) E.

67) Differentiates into antibody-producing plasma cell.

68) Causes lysis and death of foreign cells.

69) Cooperates with B cells to amplify antibody production.

70) Produces and secretes antibodies.

A. Cytotoxic T cell
B. Helper T cell
C. Memory T cell
D. B cell
E. Plasma cell

Answers: 67) D. 68) A. 69) B. 70) E.

SHORT ANSWER. Write the word or phrase that best completes the statement or answers the question.

71) The first antibody to appear in an infection belongs to the immunoglobulin class _____.

Answer: IgM

72) The process by which a microbe is coated with antibodies that make it more susceptible to phagocytosis is called _____.

Answer: opsonization

73) The only antibodies that can cross the placenta belong to the immunoglobulin class _____.

Answer: IgG

74) Antibodies that circulate bound to basophils and mast cells belong to the immunoglobulin class _____.

Answer: IgE

75) The amount of antibody in serum is called the antibody _____.

Answer: titer

76) The process of negative selection weeds out self-reactive T cells via _____, in which self-reactive T cells undergo apoptosis, or via _____, in which they remain alive but unresponsive to antigenic stimulation.

Answer: deletion; anergy

77) The three types of antigen-presenting cells include the _____, the _____, and the _____.

Answer: macrophage; B cell; dendritic cell

78) Specialized lymph capillaries in the small intestine are called _____, and the lipid-rich lymph they transport is called _____.

Answer: lacteals; chyle

79) The largest lymphatic vessel is the _____, which begins as a dilation called the _____.

Answer: thoracic duct; cisterna chyli

80) The thymus gland is located posterior to the _____.

Answer: sternum

81) The reticular cells of the germinal centers of lymph nodes are antigen-presenting cells known as _____.

Answer: dendritic cells

82) The spleen consists of two tissue types—white pulp, which is _____ , and red pulp, which is _____.

Answer: lymphatic tissue; venous sinuses (and splenic cords)

83) _____ are chemicals produced by lymphocytes, macrophages, and fibroblasts to stop viral replication in neighboring cells.

Answer: Interferons

84) Transferrins inhibit the growth of certain bacteria by reducing the amount of available _____.

Answer: iron

85) Histiocytes in the skin and microglia in the brain are examples of a group of cells called _____.

Answer: fixed macrophages

86) The four characteristic (cardinal) signs of inflammation are _____ , _____ , _____ , and _____.

Answer: heat; redness; swelling; pain

87) _____ are substances that are recognized as foreign and provoke an immune response.

Answer: Antigens

88) In cell-mediated immunity, the invading antigen is destroyed by _____.

Answer: cytotoxic T cells

89) Cell-mediated immunity is especially effective against _____ , _____ , and _____.

Answer: intracellular pathogens; certain cancer cells; foreign tissue transplants

90) Antibody-mediated immunity is especially effective against _____ and _____.

Answer: antigens in body fluids; extracellular pathogens

91) Antigens have two important characteristics—_____ , which is the ability to provoke an immune response, and _____ , which is the ability to react specifically with the antibodies or cells provoked.

Answer: immunogenicity; reactivity

92) The specific portion of an antigen that triggers an immune response is called a(n) _____.

Answer: epitope (antigenic determinant)

93) The three main types of differentiated T cells are _____ T cells that display CD4 proteins, _____ T cells that display CD8 proteins, and _____ T cells that are members of a clone that remain after an immune response.

Answer: helper; cytotoxic; memory

94) Killer T cells destroy antigens via _____, which forms holes in target cell membranes, and _____ , which activates DNA fragmenting enzymes in target cells.

Answer: perforin; lymphotoxin

95) Differentiated B cells that secrete antibodies are _____.

Answer: plasma cells

ESSAY. Write your answer in the space provided or on a separate sheet of paper.

96) Identify the components of the lymphatic system, and describe the functions of the lymphatic system.

Answer: Components include lymph, lymphatic vessels, and lymphatic tissue (nodes, spleen, thymus, red bone marrow, MALT). Functions include returning lost fluid and proteins to blood plasma to maintain BHP and BCOP, transport of dietary fats and lipid-soluble vitamins from the GI tract to the blood, and protection via specific and nonspecific immunity.

97) What is the difference between fever and the heat that is one of the characteristic signs of inflammation?

Answer: The heat of inflammation is a local event related to vasodilation induced by histamine. Greater blood flow means more heat is distributed to affected areas. Fever is a systemic increase in body temperature triggered by bacterial products and endogenous cytokines that reset the hypothalamic thermostat.

98) Describe the immunological mechanism by which anaphylactic shock occurs. Why is epinephrine useful in reversing the effects on the cardiovascular system?

Answer: IgE made in a primary response to an allergen binds to the surface of mast cells and basophils. In the secondary response, the allergen binds to the IgE and causes release of histamine (etc.) from mast cells and basophils. The result is increased vasodilation, increased capillary permeability, increased smooth muscle contraction in the bronchioles, and increased mucus secretion. Loss of fluid volume to the interstitial spaces leads to shock. Epinephrine increases the force of contraction of the heart to increase cardiac output and blood pressure and counteracts the vasodilation.

99) List the four cardinal signs of inflammation, and describe how each develops. What benefit is derived from the development of each sign?

Answer: Pain is stimulated mainly by kinins and makes a person aware of damage. Swelling results when histamine increases capillary permeability and water follows escaping proteins. This allows protective proteins and cells to reach the site. Redness and heat result as histamine increases vasodilation, which increases blood flow to the area. Increased blood flow brings needed oxygen and nutrients to cells and removes wastes. Increased temperature increases the rate of reactions and the activity of phagocytes.

100) Every cell in the body possesses molecules that could act as antigens in others. How does your own immune system "learn" not to attack your own antigens?

Answer: Some immature T cells in the thymus undergo positive selection when they become able to recognize self-MHC molecules. Those that cannot recognize MHC molecules die. Those that survive undergo negative selection, in which those cells that recognize fragments of self antigens are eliminated (deletion) or inactivated (anergy) by failure of costimulation. Similar selection of B cells occurs in red bone marrow and peripheral tissues.

CHAPTER 23　　The Respiratory System

MULTIPLE CHOICE. Choose the one alternative that best completes the statement or answers the question.

1) Which of the following is NOT considered a function of the respiratory system?
   A) regulation of acid-base balance
   B) production of red blood cells
   C) filtering inspired air
   D) transport of oxygen and carbon dioxide to tissue cells
   E) intake of oxygen and elimination of carbon dioxide

   Answer: D

2) Which of the following is NOT a function of the nose?
   A) warming of incoming air
   B) acting as a resonating chamber for speech
   C) filtering incoming air
   D) detecting olfactory stimuli
   E) gas exchange

   Answer: E

3) Which of the following lists the structures in the correct order of air flow?
   A) trachea, laryngopharynx, nasopharynx, oropharynx, larynx
   B) nasopharynx, oropharynx, laryngopharynx, trachea, larynx
   C) nasopharynx, oropharynx, laryngopharynx, larynx, trachea
   D) oropharynx, laryngopharynx, nasopharynx, larynx, trachea
   E) nasopharynx, laryngopharynx, oropharynx, larynx, trachea

   Answer: C

4) Air pressure in the middle ear is equalized via the auditory tube, which opens into the
   A) nasal cavity.
   B) maxillary sinus.
   C) nasopharynx.
   D) oropharynx.
   E) laryngopharynx.

   Answer: C

5) The vocal folds are part of the
   A) nasal cavity.
   B) laryngopharynx.
   C) trachea.
   D) larynx.
   E) lungs.

   Answer: D

6) The function of the epiglottis is to
   A) hold the pharynx open during speech.
   B) produce surfactant.
   C) close off the nasal cavity during swallowing.
   D) close off the larynx during swallowing.
   E) vibrate to produce sound as air passes over it.

   Answer: D

7) The trachea extends from the
   A) larynx to vertebra T5.
   B) soft palate to the hyoid bone.
   C) atlas to vertebra C7.
   D) epiglottis to the thyroid cartilage.
   E) foramen magnum to vertebra C5.

   Answer: A

8) C-shaped cartilage rings support the
   A) laryngopharynx.
   B) larynx.
   C) trachea.
   D) tertiary bronchi.
   E) All of these are supported by C-shaped rings.

   Answer: C

9) What is the anatomic name of the structure known as the Adam's apple?
   A) cricoid cartilage
   B) epiglottis
   C) fauces
   D) pharynx
   E) thyroid cartilage

   Answer: E.

10) The trachea and bronchi
   A) move as a result of impulses arising in the cerebrum.
   B) have "rings" of cartilage in their walls.
   C) are collapsed during a normal breathing cycle.
   D) move partially by peristalsis.
   E) help to lower the diaphragm during inhalation.

   Answer: B

11) The cough reflex is triggered by irritation of an important medical landmark called the
   A) carina.
   B) fauces.
   C) cardiac notch.
   D) cricoid cartilage.
   E) internal choanae.

   Answer: A

12) Place the following structures of the respiratory tree in order, considering how air enters the tree:
   1) secondary bronchi
   2) bronchioles
   3) alveolar ducts
   4) primary bronchi
   5) respiratory bronchioles
   6) alveoli
   7) terminal bronchioles
   A) 2, 4, 1, 7, 5, 3, 6.
   B) 4, 1, 2, 7, 5, 3, 6.
   C) 1, 4, 3, 5, 6, 3, 6.
   D) 2, 4, 7, 3, 6, 1, 5.
   E) 1, 2, 3, 4, 5, 6, 7.

   Answer: B

13) The roof of the nasal cavity is formed by the
   A) superior concha.
   B) middle concha.
   C) ethmoid bone.
   D) nasal bones.
   E) frontal bones.

   Answer: C

14) What is normally found between the visceral and parietal layers of the pleura?
    A) the lungs
    B) venous blood
    C) serous fluid
    D) air
    E) lymph

    Answer: C

15) A function of type II alveolar cells is to
    A) help control what passes between squamous epithelial cells of the alveoli.
    B) produce surfactant.
    C) act as phagocytes.
    D) produce mucus in the upper respiratory tract.
    E) store oxygen until it can be transported into the blood.

    Answer: B

16) The respiratory membrane through which gases diffuse includes ALL of the following EXCEPT
    A) type I alveolar cells.
    B) type II alveolar cells.
    C) capillary endothelium.
    D) an epithelial basement membrane.
    E) a layer of smooth muscle.

    Answer: E

17) Several small alveoli merge to form one single, larger air space. This results in a(n)
    A) increased rate of gas exchange due to an increased volume of air within the alveolus.
    B) increased rate of gas exchange due to increased partial pressure of oxygen and decreased partial pressure of carbon dioxide within the alveolus.
    C) decreased rate of gas exchange due to decreased partial pressure of oxygen and decreased partial pressure of carbon dioxide within the alveolus.
    D) decreased rate of gas exchange due to a decrease in surface area.
    E) decreased rate of gas exchange due to an increase in the thickness of the respiratory membrane.

    Answer: D

18) Boyle's law states that
    A) at a constant temperature, the volume of a gas varies inversely with the pressure.
    B) at a constant pressure, the volume of a gas is directly proportional to the temperature.
    C) the rate of diffusion is directly proportional to the surface area of the membrane.
    D) in a mixture of gases each gas exerts its own partial pressure.
    E) at a constant temperature, the volume of a gas is directly proportional to the pressure.

    Answer: A

19) During normal resting pulmonary ventilation, ALL of the following are TRUE EXCEPT
   A) the phrenic nerve stimulates contraction of the diaphragm.
   B) intrapleural pressure increases above atmospheric pressure during exhalation.
   C) air comes in during inspiration because alveolar pressure falls below atmospheric pressure.
   D) thoracic volume increases as the diaphragm contracts during inspiration.
   E) the diaphragm forms a dome as it relaxes.
   Answer: B

20) Which of the following muscles helps increase the size of the thoracic cavity during forced inspiration?
   A) external oblique
   B) external intercostals
   C) internal oblique
   D) internal intercostals
   E) pectoralis major
   Answer: B

21) Airway resistance is affected primarily by the
   A) amount of surfactant.
   B) thickness of the cartilage in the bronchial wall.
   C) amount of elastic tissue in the lungs.
   D) diameter of the bronchioles.
   E) partial pressure of each type of gas in inspired air.
   Answer: D

22) Surface tension exists in alveoli because
   A) surfactant is very sticky.
   B) elastic fibers in the basement membrane form linkages that collapse alveoli.
   C) movement of gas molecules within alveoli creates electrical charges that attract each other.
   D) polar water molecules are more strongly attracted to each other than to gas molecules in the air.
   E) polar water molecules are more strongly attracted to gas molecules in the air than to each other.
   Answer: D

23) Compliance is affected primarily by the amount of elastic tissue in the lungs and the
   A) amount of surfactant.
   B) thickness of the cartilage in the bronchial wall.
   C) partial pressure of oxygen in inspired air.
   D) diameter of the bronchioles.
   E) temperature of inspired air.
   Answer: A

24) When the diaphragm contracts,
    A) the size of the chest cavity increases.
    B) the lungs expand to fill the extra space in the chest cavity.
    C) air from outside rushes into the lungs.
    D) intrathoracic pressure decreases.
    E) all answers above are correct.

Answer: E

25) The normal resting minute volume is
    A) 6 liters/minute.
    B) 500 mL/minute.
    C) 4.5 liters/minute.
    D) 1200 mL/minute.
    E) 3.6 liters/minute.

Answer: A

26) The minute ventilation volume for someone whose tidal volume equal 450 mL, whose dead space air equal 150 mL, and whose respiratory rate is 15 respirations per minute is
    A) 2250 mL/min.
    B) 4500 mL/min.
    C) 6750 mL/min.
    D) 9000 mL/min.
    E) There is not enough information to calculate the alveolar ventilation rate.

Answer: C

27) The alveolar ventilation rate for someone whose tidal volume equal 450 mL, whose dead space air equal 150 mL, and whose respiratory rate is 15 respirations per minute is
    A) 2250 mL/min.
    B) 4500 mL/min.
    C) 6750 mL/min.
    D) 9000 mL/min.
    E) There is not enough information to calculate the alveolar ventilation rate.

Answer: B

28) The residual volume is the amount of air
    A) remaining in the lungs after the lungs collapse.
    B) that can be inhaled above tidal volume.
    C) remaining in the lungs after forced expiration.
    D) contained in air spaces above the alveoli.
    E) that can be exhaled above tidal volume.

Answer: C

29) The tidal volume is the
   A) volume of air the lungs can hold when maximally inflated.
   B) volume of air moved in and out of the lungs in a single quiet breath.
   C) percentage of alveolar air that is water vapor.
   D) sum of the inspiratory and expiratory reserve volumes.
   E) volume of air left in the lungs after a forced expiration.

   Answer: B

30) Which of the following refers to the amount of air that can be maximally inspired after a maximal expiration
   A) expiratory reserve volume.
   B) inspiratory reserve volume.
   C) inspiratory capacity.
   D) vital capacity.
   E) All answers are correct.

   Answer: D

31) On a very humid day, people with chronic respiratory diseases may experience greater difficulty breathing because
   A) they are dehydrated.
   B) atmospheric pressure is much lower, so respiratory gradients are decreased.
   C) water vapor contributes a greater partial pressure to inhaled air, thus interfering with normal gradients of other respiratory gases.
   D) the water vapor condenses within the alveoli.
   E) the water vapor decreases the solubility of oxygen.

   Answer: C

32) Dalton's Law states that
   A) at a constant temperature, the volume of a gas varies inversely with the pressure.
   B) at a constant pressure, the volume of a gas is directly proportional to the temperature.
   C) the rate of diffusion is directly proportional to the surface area of the membrane.
   D) in a mixture of gases each gas exerts its own partial pressure.
   E) at a constant temperature, the volume of a gas is directly proportional to the pressure.

   Answer: D

33) Where would you expect to find the highest partial pressure of carbon dioxide?
   A) in the atmosphere
   B) in pulmonary arteries
   C) in pulmonary veins
   D) in alveolar air
   E) in the intracellular fluid

   Answer: E

34) The reason the gradients for carbon dioxide can be smaller than those for oxygen and still meet the body's gas exchange needs is that
   A) carbon dioxide is a smaller molecule than oxygen.
   B) carbon dioxide is more water-soluble than oxygen.
   C) carbon dioxide receives assistance crossing membranes from a carrier molecule.
   D) much of the oxygen, but not the carbon dioxide, is consumed by red blood cells during transport.
   E) oxygen forms ions once it enters the alveoli, and the electrical charges slow its movement across membranes.

   Answer: B

35) Expired air has a greater oxygen content than alveolar air because
   A) more oxygen diffuses in across the mucosa of the bronchioles and bronchi.
   B) newly inspired air is entering as expired as it is leaving.
   C) oxygen is being generated by microbes in the upper respiratory tract.
   D) alveolar air mixes with air in the anatomic dead space on its way out.
   E) some carbon dioxide is converted to oxygen in respiratory passages.

   Answer: D

36) Most oxygen is transported in blood by
   A) the heme portion of hemoglobin.
   B) the globin portion of hemoglobin.
   C) simply dissolving in plasma.
   D) conversion to bicarbonate ion.
   E) any type of plasma protein.

   Answer: A

37) You would expect the partial pressure of oxygen to be highest in the
   A) pulmonary arteries.
   B) pulmonary veins.
   C) hepatic portal vein.
   D) intracellular fluid.
   E) interstitial fluid.

   Answer: B

38) If the pH of blood and interstitial fluid rises within homeostatic range, then
   A) more oxygen can combine with hemoglobin.
   B) less oxygen can stay attached to hemoglobin.
   C) the level of hydrogen ions in these fluids has increased.
   D) the increase was caused by an elevated partial pressure of carbon dioxide.
   E) the respiratory rate will increase to compensate.

   Answer: A

39) To say that hemoglobin is fully saturated means that
   A) the red blood cells contain as many hemoglobin molecules as possible.
   B) oxygen is attached to both the heme and the globin portions of the molecule.
   C) it is carrying both oxygen and carbon dioxide simultaneously.
   D) some molecule other than oxygen is attached to the oxygen-binding sites on hemoglobin.
   E) there is an oxygen molecule attached to each of the four heme groups.

   Answer: E

40) Which of the following would be TRUE if the oxygen-hemoglobin dissociation curve is shifted to the right?
   A) Partial pressure of carbon dioxide is increased.
   B) pH is increased.
   C) Temperature is decreased.
   D) Levels of BPG are decreased.
   E) Partial pressure of oxygen is decreased.

   Answer: A

41) In metabolically active tissues, you would expect
   A) the percent saturation of hemoglobin will be less than it is near the lungs.
   B) the partial pressure of oxygen will be higher than in the alveoli.
   C) the pH will be slightly higher than it is in the fluid close to the lungs.
   D) the partial pressure of carbon dioxide will be at its lowest point.
   E) All of these are correct.

   Answer: A

42) If the partial pressure of carbon dioxide is decreasing, then
    A) the partial pressure of oxygen must be increasing.
    B) the pH will also be decreasing.
    C) the affinity of hemoglobin for oxygen is decreasing.
    D) there is an increase in the rate of the reaction converting carbonic acid into water and carbon dioxide.
    E) there is an increase in the rate of the reaction converting carbonic acid into hydrogen ion and bicarbonate ion.

    Answer: D

43) BPG is a substance that
    A) is responsible for the detergent activity of surfactant.
    B) catalyzes the conversion of carbon dioxide to bicarbonate ion.
    C) is produced during glycolysis in erythrocytes and increases the dissociation of oxygen from hemoglobin.
    D) inhibits the activity of the central chemoreceptors to prolong inspiration.
    E) binds extra oxygen onto fetal hemoglobin.

    Answer: C

44) Hemoglobin will tend to bind more oxygen at a given partial pressure of oxygen if
    A) the partial pressure of carbon dioxide is increased.
    B) the temperature is increased.
    C) the pH is increased.
    D) BPG concentration increases.
    E) the concentration of hydrogen ions increases.

    Answer: C

45) High partial pressure of carbon dioxide favors the formation of
    A) BPG.
    B) carbaminohemoglobin.
    C) chloride ions.
    D) oxyhemoglobin.
    E) carbon monoxide.

    Answer: B

46) Carbonic acid is produced when

    A) oxygen combines with bicarbonate ion.
    B) carbon dioxide combines with bicarbonate ion.
    C) carbon dioxide combines with water.
    D) oxygen and carbon dioxide combine.
    E) carbon dioxide attached to hemoglobin.

    Answer: C

47) Most carbon dioxide is transported in blood by

    A) the heme portion of hemoglobin.
    B) the globin portion of hemoglobin.
    C) simply dissolving in plasma.
    D) conversion to bicarbonate ion.
    E) any plasma protein.

    Answer: D

48) The basic pattern of breathing is set by nuclei of neurons located in the

    A) pons.
    B) diaphragm.
    C) medulla oblongata.
    D) lungs.
    E) thoracic region of the spinal cord.

    Answer: C

49) The apneustic and pneumotaxic areas are located in the

    A) pons.
    B) diaphragm.
    C) medulla oblongata.
    D) lungs.
    E) thoracic region of the spinal cord.

    Answer: A

50) An enzyme that speeds the reaction of carbon dioxide and water is

    A) carbonic decarboxylase.
    B) carbonic carboxylase.
    C) carbonic anhydrase.
    D) carbonic dehydrogenase.

    Answer: C

MATCHING. Choose the item in column 2 that best matches each item in column 1.

51) Vital capacity.
52) Tidal volume.
53) Total lung volume.
54) Functional residual capacity.

   A. 500 mL
   B. 600 mL
   C. 1200 mL
   D. 4800 mL
   E. 2400 mL

Answers: 51) D. 52) A. 53) B. 54) E.

55) Tidal volume $\times$ respirations per minute.
56) Expiratory reserve capacity + residual volume.
57) (tidal volume – anatomic dead space). $\times$ respirations per minute.
58) Tidal volume+inspiratory reserve volume +expiratory reserve volume.

   A. Minute respiratory volume
   B. Functional residual volume
   C. Alveolar ventilation rate
   D. Vital capacity
   E. Tidal Volume

Answers: 55) A. 56) B. 57) C. 58) D.

59) Pressure of a gas in a closed container is inversely proportional to volume in container.
60) Each gas in a mixture of gases exerts its own pressure as if no other gases were present.
61) Quantity of a gas that will dissolve in a liquid is proportional to the partial pressure of the gas.
62) The amount of CO_2 that can be transported in the blood is influenced by the percent saturation of hemoglobin with oxygen.

   A. Henry's Law
   B. Boyle's Law
   C. Dalton's Law
   D. Haldane Effect
   E. Bohr Effect

Answers: 59) B. 60) C. 61) A. 62) D.

63) Establishes the basic rhythm of breathing.
64) Prolongs inspiration.
65) Transmits inhibitory impulses to inspiratory area.
66) Located in the pons.

   A. Expiratory area
   B. Apneustic area
   C. Pneumotaxic area
   D. Inspiratory area
   E. Answers B and C

Answers: 63) D. 64) B. 65) C. 66) E.

67) Caused by a low PO2 in arterial blood.
68) Too little hemoglobin present in blood.
69) Cyanide poisoning causes.
70) Reduced flow of blood to a given area of body.

A. Histotoxic hypoxia
B. Ischemic hypoxia
C. Anemic hypoxia
D. Anoxic anoxia
E. All answers are correct.

Answers: 67) D. 68) C. 69) A. 70) B.

SHORT ANSWER. Write the word or phrase that best completes the statement or answers the question.

71) Proteins that bind carbon dioxide form _____ compounds.

Answer: carbamino

72) The enzyme that catalyzes the conversion of carbon dioxide and water to carbonic acid is _____.

Answer: carbonic anhydrase

73) The Bohr Effect occurs when the oxygen saturation of hemoglobin decreases in response to _____.

Answer: decreased pH

74) The areas of the brain stem that make up the respiratory center are the _____ , the _____ , and the _____.

Answer: medullary rhythmicity center; apneustic center; pneumotaxic center

75) Central chemoreceptors are located in the _____ , and peripheral chemoreceptors are located in the _____ and the _____.

Answer: medulla oblongata; aortic bodies; carotid bodies

76) If hypercapnia occurs, chemoreceptors trigger a(n) _____ in the rate and depth of breathing, which is called _____.

Answer: increase; hyperventilation

77) Deficiency of hemoglobin leads to _____ hypoxia, while reduction in blood flow leads to _____ hypoxia.

Answer: anemic; ischemic

78) The _____ is a tube extending from the internal nares to the level of the cricoid cartilage.

Answer: pharynx

79) The single opening to the oropharynx is the _____.

Answer: fauces

80) The space between the vocal folds is called the _____.

Answer: rima glottidis

81) The _____ cartilage of the larynx is the landmark for making an emergency airway.

Answer: cricoid

82) The pitch of a sound generated by the larynx is controlled by the _____.

Answer: tension on the vocal cords

83) The epithelial portion of the mucosa of the trachea is _____ epithelium.

Answer: pseudostratified ciliated columnar

84) An aspirated object is more likely to lodge in the _____ primary bronchus due to its structure.

Answer: right

85) Tertiary bronchi branch to form _____.

Answer: bronchioles

86) Binding of an antigen to IgE antibodies could result in _____ of bronchioles due to release of _____.

Answer: constriction; histamine

87) The concavity on the medial surface of the left lung is called the _____.

Answer: cardiac notch

88) The function of type I alveolar cells is _____ ; the function of type II alveolar cells is _____.

Answer: gas exchange; secretion of alveolar fluid

89) The detergent component of alveolar fluid is called _____ , the function of which is to _____.

Answer: surfactant; lower surface tension of alveolar fluid

90) The components of the respiratory membrane include the _____ , the _____ , the _____ , and the _____.

Answer: alveolar wall (Type I and II cells); epithelial basement membrane of the alveolar wall; basement membrane of the capillary; endothelium of the capillary

91) Pulmonary capillaries _____ in response to hypoxia.

Answer: constrict

92) Thoracic volume decreases when the diaphragm _____ , causing alveolar pressure to _____ according to _____ Law.

Answer: relaxes; increase; Boyle's

93) Entry of air into the intrapleural space is called _____ .

Answer: pneumothorax

94) Forced expiration against a closed rima glottidis is called _____ .

Answer: the Valsalva maneuver

95) Increased mucus production by the mucosa of the bronchial tree interferes with ventilation because it increases _____ .

Answer: airway resistance

ESSAY. Write your answer in the space provided or on a separate sheet of paper.

96) Describe the inward forces of elastic recoil, and explain why the lungs do not normally collapse during expiration.

Answer: Elastic recoil is the recoil of elastic fibers stretched during inspiration and the pull of the surface tension of alveolar fluid. Intrapleural pressure is always subatmospheric during normal breathing, which tends to pull lungs outward and to keep alveolar pressure from equalizing with atmospheric pressure. Surfactant in alveolar fluid decreases surface tension to help prevent collapse.

97) Why is epinephrine injected as a treatment for the respiratory signs and symptoms of anaphylaxis?

Answer: Epinephrine enhances sympathetic activity to dilate airways and decrease airway resistance, which had been elevated by the effects of histamine on the bronchioles. It also raises blood pressure, which enhances oxygen delivery to tissues by increasing flow.

98) Describe and explain the effects of smoking on the functioning of the respiratory system.

Answer: Nicotine constricts terminal bronchioles to increase airway resistance, as do the increased mucus secretion and swelling of the mucosa. Smoke inhibits the movement of cilia, which allows the buildup of substances and microbes normally removed. Over time, smoking leads to destruction of elastic tissue, which decreases compliance, and ultimately to the effects of emphysema.

99) Describe the neural, chemical, and physical changes that increase the rate and depth of ventilation during exercise.

Answer: Anticipation of exercise generates neural input to the limbic system. Sensory input is provided from proprioceptors, and motor input is provided from the primary motor cortex. As the partial pressure of oxygen falls due to increased consumption, the partial pressure of carbon dioxide and the temperature increase due to metabolic activity in muscle fibers. Also, carbon dioxide is added via the buffering of the hydrogen ions produced as a result of lactic acid production. Chemoreceptors sense the changes in partial pressure and notify the medullary rhythmicity center to increase the rate and depth of breathing.

100) In chronic emphysema, some alveoli merge together and some are replaced with fibrous connective tissue. In addition, the bronchioles are often inflamed, and expiratory volume is reduced. Using proper respiratory system terminology, explain at least four reasons why affected individuals will have problems with ventilation and external respiration.

Answer: Answers could include: reduced compliance (reduces ability to increase thoracic volume); increased airway resistance (decreases tidal volume); decreased diffusion due to increased diffusion distance, decreased surface area, and changes in partial pressures of gases (altering gradients). Other answers may be acceptable.

CHAPTER 24   The Digestive system

MULTIPLE CHOICE. Choose the one alternative that best completes the statement or answers the question.

1) The type of chemical reaction catalyzed by the digestive enzymes in the digestive juices of the alimentary canal is
   A) oxidation.
   B) reduction.
   C) hydrolysis.
   D) dehydration.
   E) phosphorylation.

   Answer: C

2) Which of the following would be considered an accessory organ of the digestive system?
   A) pancreas.
   B) stomach.
   C) esophagus.
   D) large intestine.
   E) small intestine.

   Answer: A

3) Which of the following lists the tubing in the correct order of food movement?
   A) nasopharynx, oropharynx, laryngopharynx, larynx, esophagus
   B) oropharynx, laryngopharynx, esophagus, stomach, pyloric valve
   C) laryngopharynx, oropharynx, esophagus, stomach, pyloric valve
   D) oropharynx, laryngopharynx, esophagus, pyloric valve, stomach
   E) nasopharynx, oropharynx, larynx, esophagus, stomach

   Answer: B

4) The regular contractions of the muscularis that push food through the entire gastrointestinal tract are known as
   A) segmentations.
   B) haustral churning.
   C) peristalsis.
   D) pendular movements.
   E) migratory motility complex.

   Answer: C

5) Peristalsis occurs during
   A) the voluntary stage of deglutition.
   B) the pharyngeal stage of deglutition.
   C) the esophageal stage of deglutition.
   D) mastication.
   E) Both B and C are correct.

   Answer: C

6) The muscularis of most organs of the gastrointestinal tract consists of two layers of smooth muscle EXCEPT in the
   A) duodenum.
   B) ileum.
   C) sigmoid colon.
   D) stomach.
   E) ascending colon.

   Answer: D

7) Intrinsic factor secreted by parietal cells of the stomach is required for
   A) activation of pepsin.
   B) buffering of HCl.
   C) complete gastric emptying.
   D) absorption of vitamin B_{12}.
   E) stimulation of mixing waves.

   Answer: D

8) The folds of the gastric mucosa are called
   A) microvilli.
   B) circular folds.
   C) gastric pits.
   D) villi.
   E) rugae.

   Answer: E

9) The pyloric sphincter is located at the junction of the
   A) esophagus and stomach.
   B) stomach and duodenum.
   C) ileum and cecum.
   D) esophagus and larynx.
   E) sigmoid colon and rectum.

   Answer: B

10) Bicarbonate ions diffuse into blood capillaries of the stomach after a meal because
    A) they are being generated from amino acids absorbed by the gastric mucosa.
    B) they are being exchanged for hydrogen ions that enter the stomach lumen.
    C) they are being exchanged for chloride ions that enter the stomach lumen.
    D) they are being exchanged for potassium ions that enter the stomach lumen.
    E) carbon dioxide is generated as pepsin hydrolyzes proteins, and it is converted to bicarbonate ion.

Answer: C

11) Which of the following has the lowest pH?
    A) saliva.
    B) gastric juice.
    C) pancreatic juice.
    D) bile.
    E) intestinal juice.

Answer: B

12) Which of the following occurs during the cephalic phase of gastric digestion?
    A) Chemoreceptors detect change in the pH of gastric juice.
    B) Stretch receptors detect distention of the stomach.
    C) Chemoreceptors detect fatty acids in the duodenum.
    D) Sight, smell, thought, or taste of food triggers parasympathetic impulses.
    E) CCK is secreted by enteroendocrine cells.

Answer: D

13) Gastric emptying is stimulated by ALL of the following EXCEPT
    A) distention of the stomach.
    B) gastrin.
    C) CCK.
    D) partially digested proteins.
    E) the vagus nerve.

Answer: C

14) Salivary glands include
    A) parotid glands.
    B) submandibular glands.
    C) sublingual glands.
    D) All answers are correct.

Answer: D

15) A hormone produced by cells of the pyloric portion of the stomach is
    A) secretin.
    B) enterogastrone.
    C) insulin.
    D) gastrin.
    E) glucagon.

    Answer: D

16) Salivary secretions
    A) aid in chewing and swallowing.
    B) initiate digestion of starches.
    C) are needed for tasting.
    D) moisten and lubricate food.
    E) function in all of the above ways.

    Answer: E

17) An enzyme that digests carbohydrates is
    A) trypsin.
    B) pepsin.
    C) mucin.
    D) lipase.
    E) amylase.

    Answer: E

18) The salivary enzyme amylase (ptylin) functions to digest
    A) proteins.
    B) fats.
    C) carbohydrates.
    D) proteins.
    E) nucleic acids.

    Answer: C

19) The primary digestive function of the stomach is
    A) carbolytic.
    B) lipolytic.
    C) proteolytic.
    D) saccarolytic.
    E) All of the answers are correct.

    Answer: C

20) In areas of the gastrointestinal tract specialized for absorption of nutrients, the type of epithelium seen in the mucosa is
    A) simple squamous.
    B) stratified squamous.
    C) transitional.
    D) simple columnar.
    E) pseudostratified ciliated columnar.

    Answer: D

21) The small intestine is attached to the posterior abdominal wall by a fold of the peritoneum called the
    A) mesocolon.
    B) mesentery.
    C) falciform ligament.
    D) taeniae coli.
    E) greater omentum.

    Answer: B

22) Which of the following lists the tubing in the correct order of food movement?
    A) pyloric valve, duodenum, ileum, jejunum, ileocecal valve
    B) pyloric valve, jejunum, duodenum, ileum, ileocecal valve
    C) ileocecal valve, ileum, jejunum, duodenum, pyloric valve
    D) ileocecal valve, ileum, jejunum, duodenum, pyloric valve
    E) pyloric valve, duodenum, jejunum, ileum, ileocecal valve

    Answer: E

23) During swallowing, the nasal cavity is closed off by the soft palate and the
    A) epiglottis.
    B) uvula.
    C) palatine tonsils.
    D) fauces.
    E) tongue.

    Answer: B

24) The process of mastication results in
    A) passage of food from the oral cavity into the esophagus.
    B) removal of pathogens from partially digested food by MALT tissues.
    C) mechanical mixing of food with saliva and shaping of food into a bolus.
    D) sudden movement of colonic contents into the rectum.
    E) passage of feces from the anus.

    Answer: C

25) Which of the following is an example of mechanical digestion?

A) glycolysis

B) defecation

C) oxidation-reduction

D) mastication

E) hydrolysis

Answer: D

26) Partially digested food is usually passed from the stomach to the small intestine about how long after consumption?

A) an hour or less

B) 2–4 hours

C) 6–8 hours

D) 10–12 hours

E) 24 hours

Answer: B

27) Increased activity of the sympathetic nervous system will

A) increase production of all hydrolytic enzymes by abdominal organs.

B) increase only production of those digestive juices rich in buffers.

C) have no effect on the digestive system.

D) decrease production of digestive juices.

E) increase movement of food through the alimentary canal.

Answer: D

28) Which of the following is NOT produced by the acini of the pancreas?

A) amylase

B) lipase

C) carboxypeptidase

D) somatostatin

E) elastase

Answer: D

29) The functions of the gallbladder include

A) production of bile.

B) storage and concentration of bile.

C) formation of urea.

D) secretin of cholecystokinin.

E) Both A and B are correct.

Answer: B

30) The function of bile is to
   A) emulsify fats.
   B) transport fats through the blood.
   C) hydrolyze fats.
   D) actively transport fats through epithelial membranes.
   E) All of these are correct.

   Answer: A

31) The common bile duct is formed by the union of the
   A) right and left hepatic ducts.
   B) cystic and pancreatic ducts.
   C) common hepatic and cystic ducts.
   D) all bile capillaries.
   E) pancreatic and accessory ducts.

   Answer: C

32) The greenish color of bile is the result of the presence of breakdown products of
   A) hemoglobin.
   B) urea.
   C) starch.
   D) the B vitamins.
   E) fats.

   Answer: A

33) Gallstones are usually made of crystallized
   A) glucose.
   B) bilirubin.
   C) cholesterol.
   D) chyme.
   E) fat-soluble vitamins.

   Answer: C

34) Without functioning hepatocytes, protein catabolism is a toxic process due to
   A) production of ammonia.
   B) excessive HCl production.
   C) buildup of the parts of amino acids remaining after deamination.
   D) clogging of bile canaliculi with denatured proteins.
   E) increased fluid volume.

   Answer: A

35) The liver produces urea to
   A) detoxify ammonia produced via deamination of proteins.
   B) keep bile in an inactive form until it reaches the small intestine.
   C) convert into glucose when blood glucose is low.
   D) store iron.
   E) bind to ingested poisons to detoxify them.

   Answer: A

36) The major stimulus for secretion of secretin is
   A) the sight and aroma of food.
   B) entry of a bolus into the esophagus.
   C) CCK.
   D) distention of the stomach.
   E) entry of acid chyme into the small intestine.

   Answer: E

37) The difference between the effects of secretin and the effects of CCK on the pancreas is that
   A) secretin stimulates secretion of pancreatic juice, while CCK inhibits secretion of pancreatic juice.
   B) secretin stimulates the acini of the pancreas, while CCK stimulates the pancreatic islets.
   C) secretin stimulates alpha cells, while CCK stimulates beta cells.
   D) secretin causes dilation of the pancreatic duct, while CCK causes constriction of the duct.
   E) secretin stimulates secretion of pancreatic juice rich in bicarbonate, while CCK inhibits secretion of pancreatic juice rich in digestive enzymes.

   Answer: E

38) The hydrolytic reactions catalyzed by trypsin and chymotrypsin would result in the production of
   A) fatty acids and glycerol.
   B) monosaccharides.
   C) peptides.
   D) nucleotides.
   E) dextrin.

   Answer: C

39) Specific disaccharides are hydrolyzed by enzymes found in
   A) gastric juice.
   B) intestinal juice.
   C) saliva.
   D) pancreatic juice.
   E) bile.

   Answer: B

40) The products of the hydrolysis reaction catalyzed by carboxypeptidase are
    A) amino acids.
    B) glucose and fructose.
    C) dextrins.
    D) nitrogenous bases.
    E) fatty acids and monoglycerides.

    Answer: A

41) Most absorption of nutrients occurs in the
    A) mouth.
    B) transverse colon.
    C) stomach.
    D) small intestine.
    E) rectum.

    Answer: D

42) Monosaccharides enter the capillaries of the villi from epithelial cells by
    A) primary active transport.
    B) facilitated diffusion.
    C) simple diffusion.
    D) secondary active transport linked to sodium ion transport.
    E) emulsification.

    Answer: B

43) Glucose is transported into epithelial cells of the villi via
    A) secondary active transport coupled to active transport of sodium ions.
    B) secondary active transport coupled to active transport of galactose.
    C) facilitated diffusion.
    D) primary active transport.
    E) pinocytosis.

    Answer: A

44) The primary chemical digestion in the large intestine results from the action of
    A) the continued action of pancreatic juice.
    B) bacterial enzymes.
    C) bilirubin.
    D) hydrochloric acid.
    E) fat-soluble vitamins.

    Answer: B

45) The appendix is attached to the
   A) left lobe of the liver.
   B) gallbladder.
   C) cecum.
   D) rectum.
   E) splenic flexure.

   Answer: C

46) Which of the following lists the tubing in the correct order of food movement?
   A) descending colon, splenic flexure, transverse colon, hepatic flexure, ascending colon, sigmoid colon
   B) ascending colon, hepatic flexure, transverse colon, splenic flexure, descending colon, sigmoid colon
   C) sigmoid colon, ascending colon, hepatic flexure, transverse colon, splenic flexure, descending colon
   D) ascending colon, splenic flexure, transverse colon, hepatic flexure, descending colon, sigmoid colon
   E) sigmoid colon, descending colon, splenic flexure, transverse colon, hepatic flexure, ascending colon

   Answer: B

47) The large intestine absorbs mostly
   A) amino acids.
   B) monosaccharides.
   C) bile pigments.
   D) water.
   E) triglycerides.

   Answer: D

48) The normal color of feces is due primarily to the
   A) pigments in the bacteria present.
   B) breakdown products of hemoglobin.
   C) pigments in foods consumed.
   D) pigments in epithelial cells sloughed off from the mucosa.
   E) chemical interactions of undigested foods.

   Answer: B

49) Emptying of bile from the gallbladder is controlled by
   A) cholecystokinin.
   B) enterokinase.
   C) enterocrinin.
   D) gastrin.
   E) pepsin.
   Answer: A

50) The intestinal enzyme that functions to digest fat is
   A) bile.
   B) lipase.
   C) pepsin.
   D) trypsin.
   E) All answers are correct.
   Answer: B

MATCHING. Choose the item in column 2 that best matches each item in column 1.

51)	Secrete trypsin.	A.	Hepatocytes
52)	Secrete bile.	B.	Parietal cells of stomach
53)	Secrete hydrochloric acid.	C.	Parotid glands
54)	Secrete saliva.	D.	Acini of pancreas
		E.	Chief cells of stomach

Answers: 51) D. 52) A. 53) B. 54) C.

55)	Bile acts in	A.	Jejunum
56)	Connects to caecum.	B.	Ileum
57)	Appendix is attached to	C.	Duodenum
58)	Connects to sigmoid colon.	D.	Ascending colon
		E.	Descending colon

Answers: 55) C. 56) B. 57) D. 58) E.

59)	Digest lipids.	A.	Amylase
60)	Digest proteins.	B.	Trypsin
61)	Emulsifies fats.	C.	Lipases
62)	Digest carbohydrates.	D.	Bile
		E.	Secretin

Answers: 59) C. 60) B. 61) D. 62) A.

63) Acts on the stomach.
64) Acts to release alkalyn solution from pancreas.
65) Acts to release enzymes from pancreas.
66) Acts to release bile from gallbladder.

A. Secretin
B. Cholecystokinin
C. Gastrin
D. Enterogastrone
E. Glucagon

Answers: 63) C. 64) A. 65) B. 66) B

67) Connects esophagus to stomach.
68) Connects small intestine to large intestine.
69) Connects stomach to duodenum.
70) Connects rectum to the outside.

A. Pyloric valve
B. Ileocaecal valve
C. Cardiac valve
D. Anal valve
E. Aortic valve

Answers: 67) C. 68) B. 69) A. 70) D.

SHORT ANSWER. Write the word or phrase that best completes each statement or answers the question.

71) The epithelium in the mouth, pharynx, and esophagus is _____ epithelium.

Answer: nonkeratinized stratified squamous

72) The part of the enteric nervous system located in the muscularis of the wall of the gastrointestinal tract is the _____.

Answer: myenteric plexus

73) The condition in which fluid accumulates in the peritoneal cavity is called _____.

Answer: ascites

74) The largest peritoneal fold is called the _____.

Answer: greater omentum

75) The antibodies seen in saliva belong to the immunoglobulin class _____.

Answer: IgA

76) The role of chloride ions in saliva is to _____; the role of bicarbonate and phosphate ions is to _____.

Answer: activate salivary amylase; act as buffers

77) Teeth are composed primarily of _____.

Answer: dentin

78) The root of a tooth is held to the periodontal ligament by _____.

Answer: cementum

79) The process of mechanical digestion in the mouth is called _____.

Answer: mastication (chewing)

80) The opening in the diaphragm through which the esophagus passes is called the _____.

Answer: esophageal hiatus

81) Chief cells of the stomach produce _____ and _____.

Answer: pepsinogen; gastric lipase

82) Parietal cells of the stomach secrete _____ and _____.

Answer: hydrochloric acid; intrinsic factor

83) Mixing waves of the stomach convert solid food to a liquid called _____.

Answer: chyme

84) During the gastric phase of gastric digestion, acetylcholine from parasympathetic neurons stimulates secretion of the hormone _____.

Answer: gastrin

85) Once food enters the duodenum, enteroendocrine cells in the small intestine secrete _____ and _____.

Answer: CCK; secretin

86) Gastric emptying is slowest after a meal rich in _____.

Answer: lipids (triglycerides)

87) The principal triglyceride-digesting enzyme in adults is _____.

Answer: pancreatic lipase

88) Bile is secreted by hepatocytes into vessels called _____.

Answer: bile canaliculi

89) _____ is an important phospholipid in bile that helps make cholesterol more water-soluble.

Answer: Lecithin

90) The principal bile pigment is _____.

Answer: conjugated bilirubin

91) The phagocytic cells of the liver are called _____.

Answer: stellate reticuloendothelial cells

92) All of the microvilli of the epithelial cells of the small intestine collectively form a fuzzy line called the _____.

Answer: brush border

93) Aggregated lymphatic follicles (Peyer's Patches) are located in the mucosa of the _____ of the small intestine.

Answer: ileum

94) Chyme and digestive juices are mixed by localized contractions of the muscularis of the small intestine called _____.

Answer: segmentations

95) Chyme normally remains in the small intestine for about _____ hours.

Answer: 3–5

ESSAY. Write your answer in the space provided or on a separate sheet of paper.

96) Explain why food does not normally go up into your nasal cavity or down into your lungs when you swallow-even if you are standing on your head when you swallow.

Answer: The presence of food in the oropharynx stimulates the deglutition center in the medulla and pons to move the soft palate and uvula upward to close off the nasopharynx, thus keeping food out of the nasal cavity. At the same time, the larynx rises and the epiglottis moves down and back to seal off the larynx, which is further closed by the vocal cords, thus keeping food from entering the lower respiratory tract.

97) Describe the structural characteristics of the small intestine that enhance its function as the major absorber of nutrients.

Answer: All structures increase surface area to increase the rate of reabsorption: great length (10′ in living humans), microvilli on plasma membrane of each epithelial cell, villi (fingerlike projections of mucosa), and circular folds (permanent ridges in the mucosa).

98) Describe the role of the liver in protein metabolism.

Answer: Hepatocytes deaminate amino acids. The amine group is converted to toxic ammonia. Hepatocytes convert the toxic ammonia to less toxic urea for excretion in urine. The liver also synthesizes many proteins, including most plasma proteins.

99) Describe the structures and functions of the enteric nervous system.

Answer: The ENS consists of the submucosal plexus in the submucosa and the myenteric plexus in the musclaris. Both contain sensory and motor neurons, as well as ANS postganglionic fibers of both divisions. The myenteric plexus also contains parasympathetic ganglia. The submucosal plexus regulates movements of the mucosa, secretion from glands in the gastrointestinal tract, and vasoconstriction of blood vessels in the gastrointestinal tract. The myenteric plexus regulates gastric motility.

100) Identify the protein-hydrolyzing enzymes in the digestive tract, and name their sources. Why are these enzymes released in an inactive form?

Answer: Pepsin from the stomach, trypsin, chymotrypsin, carboxypeptidase, and elastin from the pancreas, and aminopeptidase and dipeptidase from the small intestine are the proteases in the GI tract. The enzymes are not activated until they are in the lumen of the stomach or small intestine because they would otherwise digest the proteins in the cells that produce them.

CHAPTER 25   Metabolism

MULTIPLE CHOICE. Choose the one alternative that best completes the statement or answers the question.

1) Most biological oxidations are
   A) dehydrogenation reactions.
   B) dehydration reactions.
   C) hydrolysis reactions.
   D) phosphorylation reactions.
   E) decarboxylation reactions.

   Answer: A

2) Conversion of NAD^+ to NADH and H^+ is an example of
   A) oxidation.
   B) reduction.
   C) phosphorylation.
   D) hydrolysis.
   E) dehydration.

   Answer: B

3) The role of insulin in the body's utilization of glucose is to
   A) catalyze the conversion of glucose 6-phosphate into glycogen.
   B) carry acetyl units into the Krebs cycle.
   C) increase the rate of facilitated diffusion of glucose into cells.
   D) transport hydrogen ions between compounds of the Krebs cycle and compounds of the electron transport chain.
   E) increase the rate of glycolysis.

   Answer: C

4) Lactic acid is produced as a result of the chemical reduction of
   A) acetyl CoA.
   B) oxaloacetic acid.
   C) pyruvic acid
   D) cytochromes.
   E) NAD.

   Answer: C

5) The function of coenzyme A in glucose metabolism is to
   A) reduce pyruvic acid to lactic acid.
   B) convert glucose 6-phosphate into glycogen.
   C) transport glucose from the blood across cell membranes.
   D) carry hydrogen ions between compounds in the Krebs cycle and compounds in the electron transport chain.
   E) carry two-carbon units into the Krebs cycle.

   Answer: E

6) The primary significance of the Krebs cycle in terms of ATP production is
   A) production of large amounts of GTP that can be converted to ATP.
   B) transfer of energy into NADH and $FADH_2$.
   C) generation of carbon dioxide for use in the electron transport chain.
   D) transfer of energy into ATP between each step of the cycle.
   E) production of acetyl units.

   Answer: B

7) Decarboxylation reactions occur in
   A) the electron transport chain.
   B) the Krebs cycle.
   C) chemiosmosis.
   D) glycogenolysis.
   E) ketogenesis.

   Answer: B

8) Each molecule of acetyl CoA that enters the Krebs cycle produces how many molecules of carbon dioxide?
   A) one.
   B) two.
   C) four.
   D) six.
   E) 36.

   Answer: B

9) The enzyme found in the hydrogen ion channels between the inner and outer mitochondrial membrane is:

   A) cytochrome oxidase.

   B) ATP synthase.

   C) NADH dehydrogenase.

   D) citric synthetase.

   E) succinyl kinase.

   Answer: B

10) The enzyme that converts pyruvate into acetyl choline is

   A) dehydrogenase.

   B) decarboxylase.

   C) carboxylase.

   D) transaminase.

   E) synthetase.

   Answer: B

11) The end-products of the complete aerobic oxidation of glucose are

   A) fatty acids and glycerol.

   B) ATP and oxygen.

   C) amino acids.

   D) carbon dioxide, water and energy.

   E) pyruvic acid and lactic acid.

   Answer: D

12) Most ATP generated by the complete oxidation of glucose results from the reactions of

   A) glycogenolysis.

   B) glycolysis.

   C) the Krebs cycle.

   D) the electron transport chain.

   E) gluconeogenesis.

   Answer: D

13) Which of the following substances increases in amount during cellular respiration?

   A) ATP

   B) glucose

   C) oxygen

   D) glycogen

   E) ADP

   Answer: A

14) Which of the following processes requires oxygen?

 A) lactic acid production

 B) glycolysis

 C) electron transport system

 D) gluconeogenesis

 E) All of the above answers are correct.

Answer: C

15) Anabolic processes

 A) are the same as catabolic processes.

 B) result, finally, in the death of the individual.

 C) are the building up or synthesizing, processes.

 D) are unaffected by hormones.

 E) are all the result of exergonic reactions.

Answer: C

16) Metabolism in living organisms is mediated by specific organic catalysts called

 A) mitochondria.

 B) hormones.

 C) vitamins.

 D) enzymes.

 E) steroids.

Answer: D

17) The most common carbohydrate "fuel" for living organisms is

 A) sucrose.

 B) maltose.

 C) glucose.

 D) fructose.

 E) lactose.

Answer: C

18) Enzymes of the Krebs cycle and electron transport chain are located in the

 A) mitochondria.

 B) endoplasmic reticulum.

 C) nucleus.

 D) lysosomes.

 E) peroxisomes.

Answer: A

19) Which of the following could be anaerobic?

A) electron transport chain

B) Krebs cycle

C) respiration

D) glycolysis

E) lipolysis

Answer: D

20) Place the following compounds in their correct sequence for glycolysis.

1   glyceraldehyde-3-phosphate

2   1,3-diphosphoglyceric acid

3   2-phosphoglyceric acid

4   phosphoenolpyruvic acid

5   fructose 1, 6-diphosphate

A) 1, 2, 3, 4, 5.

B) 2, 4, 3, 1, 5.

C) 3, 5, 1, 2, 4.

D) 5, 1, 2, 3, 4.

E) 4, 2, 1, 3, 5.

Answer: D

21) What is gluconeogenesis? It is concerned with the

A) breakdown of glucose.

B) deamination of amino acids.

C) production of glycogen.

D) ability of the liver to form glucose from noncarbohydrate substances.

E) All of the answers are correct.

Answer: D

22) Glycogenesis is the

A) process by which the liver builds glycogen from noncarbohydrate precursors.

B) breakdown of fats to fatty acids and glycerol.

C) breakdown of glycogen to reform glucose.

D) process of glycogen formation.

E) All of the above answers are correct.

Answer: D

23) For glycerol to be used in carbohydrate metabolism, it is first converted into
   A) fatty acids.
   B) oxaloacetic acid.
   C) glyceraldehyde 3-phosphate.
   D) glucose 6-phosphate.
   E) citric acid.

   Answer: C

24) Which of the following is an immediate, direct product of protein digestion?
   A) pyruvic acid
   B) amino acid
   C) ammonia
   D) carbon dioxide
   E) urea

   Answer: B

25) Gluconeogenesis occurs primarily in hepatocytes and
   A) adipocytes.
   B) neurons.
   C) skeletal muscle fibers.
   D) kidney cortex cells.
   E) cardiac muscle cells.

   Answer: D

26) For glycogen to be used for energy production, it must first be converted into
   A) glycerol.
   B) glucose 6-phosphate.
   C) carbon dioxide.
   D) lactic acid.
   E) an acetyl unit.

   Answer: B

27) Excess cholesterol is transported to the liver for elimination by
   A) high-density lipoproteins.
   B) low-density lipoproteins.
   C) very low-density lipoproteins.
   D) chylomicrons.
   E) ketone bodies.

   Answer: A

28) Cholesterol is carried to cells for repair of membranes and synthesis of steroid hormones and bile salts by
   A) high-density lipoproteins.
   B) low-density lipoproteins.
   C) very low-density lipoproteins.
   D) chylomicrons.
   E) ketone bodies.
   Answer: B

29) Endogenous triglycerides synthesized in hepatocytes are transported to adipocytes for storage by
   A) high-density lipoproteins.
   B) low-density lipoproteins.
   C) very low-density lipoproteins.
   D) chylomicrons.
   E) ketone bodies.
   Answer: C

30) Dietary lipids are transported in lymph and blood by
   A) high-density lipoproteins.
   B) low-density lipoproteins.
   C) very low-density lipoproteins.
   D) chylomicrons.
   E) ketone bodies.
   Answer: D

31) Most triglycerides are stored in adipocytes in the
   A) areas between the muscles.
   B) areas around the kidneys.
   C) subcutaneous tissue.
   D) greater omentum.
   E) large intestine.
   Answer: C

32) The processes of lipogenesis, protein synthesis, and glycogenesis, are all promoted by which of the following hormones?
   A) human growth hormone.
   B) glucagon.
   C) cortisol.
   D) insulin.
   E) epinephrine.
   Answer: D

33) Beta oxidation is the process by which
   A) hydrogen ions are removed from compounds in the Krebs cycle.
   B) carbon dioxide is removed from compounds in the Krebs cycle.
   C) amine groups are removed from proteins.
   D) fatty acids are broken down for use in the Krebs cycle.
   E) ADP is converted to ATP.

   Answer: D

34) The conversion of glycerol into glyceraldehyde 3-phosphate for use in glycolysis is an example of
   A) glycogenolysis.
   B) deamination.
   C) beta oxidation.
   D) oxidative phosphorylation.
   E) gluconeogenesis.

   Answer: E

35) A compound that is a product of ketogenesis is
   A) oxaloacetic acid.
   B) cholesterol.
   C) pyruvic acid.
   D) glucose 6-phosphate.
   E) beta-hydroxybutyric acid.

   Answer: E

36) Before amino acids can enter the Krebs cycle, they must be
   A) oxidized.
   B) reduced.
   C) decarboxylated.
   D) deaminated.
   E) dehydrated.

   Answer: D

37) The process of transamination results in
   A) synthesis of nonessential amino acids.
   B) conversion of amino acids to glucose.
   C) phosphorylation of glucose.
   D) conversion of a hexose to a pentose.
   E) production of ammonia.

   Answer: A

38) Urea is produced in the process of detoxifying
   A) ammonia.
   B) lactic acid.
   C) carbon dioxide.
   D) pyruvic acid.
   E) ketone bodies.

   Answer: A

39) The compound that is converted into urea by the liver is formed from the
   A) acetyl units formed during lipolysis.
   B) amine groups removed during deamination.
   C) reactions of the electron transport chain.
   D) reactions of glycogenolysis.
   E) lactic acid formed during anaerobic respiration.

   Answer: B

40) Conversion of amino acids to oxaloacetic acid to glucose 6-phosphate is an example of
   A) ketogenesis.
   B) beta oxidation.
   C) glycolysis.
   D) deamination.
   E) gluconeogenesis.

   Answer: E

41) The primary hormone regulating the metabolic reactions and membrane transport activities of the absorptive state is
   A) glucagon.
   B) thyroxine.
   C) epinephrine.
   D) insulin.
   E) cortisol.

   Answer: D

42) ALL of the following hormones raise blood glucose EXCEPT
   A) human growth hormone.
   B) glucagon.
   C) cortisol.
   D) insulin.
   E) epinephrine.

   Answer: D

43) The function of the satiety center is to
A) regulate the rate of lipogenesis.
B) stimulate consumption of food.
C) regulate the release of insulin.
D) stimulate cessation of feeding.
E) regulate body temperature.

Answer: D

44) The hormones primarily responsible for daily regulation of production of body heat are produced by the
A) pancreas.
B) thyroid gland.
C) adrenal cortex.
D) adrenal medulla.
E) hypothalamus.

Answer: B

45) In chemiosmosis, ATP is produced when
A) a high-energy phosphate group is passed from glucose 6-phosphate to ADP.
B) hydrogen ions are bound to NAD and FAD.
C) pyruvic acid is converted to an acetyl unit.
D) hydrogen ions diffuse into the mitochondrial matrix.
E) glucose is transported across the cell membrane.

Answer: D

46) Vasoconstriction is considered to be a heat-saving mechanism because
A) the energy released by vascular smooth muscle contraction warms you up.
B) sweat glands don't receive enough blood supply to produce sweat.
C) less heat can be conducted to the surface and radiated away.
D) more heat is generated by friction as a larger volume of blood is forced through a smaller space.
E) All of these are correct.

Answer: C

47) Dietary fat is required for the absorption of ALL of the following EXCEPT
A) vitamin A.
B) tocopherols.
C) vitamin K.
D) vitamin D.
E) thiamine.

Answer: E

48) Beta-carotene is the provitamin form of

A) vitamin A.

B) biotin.

C) vitamin B_{12}.

D) ascorbic acid.

E) tocopherols.

Answer: A

49) Sunlight converts 7-dehydrocholesterol in the skin to cholecalciferol, which is a form of

A) vitamin A.

B) niacin.

C) vitamin C.

D) vitamin D.

E) folic acid.

Answer: D

50) The complete hydrolysis of proteins yields

A) amino acids.

B) fatty acids and glycerol.

C) nucleic acids.

D) monosaccharides.

E) carbon dioxide and water.

Answer: A

MATCHING. Choose the item in column 2 that best matches each item in column 1.

51) Most abundant cation in extracellular fluid.   A. Calcium

52) Principal cation in intracellular fluid.   B. Iron

53) Binds oxygen in hemoglobin.   C. Potassium

54) Stored in bones and teeth.   D. Sodium

                                                          E. Copper

Answers: 51) D. 52) C. 53) B. 54) A.

55) Essential for synthesis of prothrombin.   A. Vitamin A

56) Essential for formation of rhodopsin.   B. Vitamin C

57) Essential for absorption and utilization   C. Vitamin D
    of calcium.   D. Vitamin K

58) Antioxidant important in wound healing.   E. Vitamin E

Answers: 55) D. 56) A. 57) C. 58) B.

59) Component in coenzymes NAD and NADP.   A.   Nicotinamide

60) Components of coenzymes FAD and FMN.   B.   Riboflavin

61) Important in red blood cell formation.   C.   Thiamin

62) Important in nucleic acid synthesis.   D.   Vitamin B_{12}

                                              E.   Folic acid

Answers: 59) A. 60) B. 61) D. 62) E.

63) Synthesis of glycogen.   A.   Glycolysis

64) Synthesis of lipids.   B.   Glycogenesis

65) Breakdown of glucose.   C.   Lipogenesis

66) Formation of glucose from other chemicals.   D.   Gluconeogenesis

                                              E.   Lipolysis

Answers: 63) B. 64) C. 65) A. 66) D.

67) Triglycerides secreted into lymph.   A.   Insulin

68) Used to form Vitamin D.   B.   Low-density lipoprotein

69) Brings cholesterol from the liver to the tissues.   C.   High-density lipoprotein

70) Promotes absorption of glucose by liver cells.   D.   Chylomicrons

                                              E.   Cholesterol

Answers: 67) D. 68) E. 69) B. 70) A.

SHORT ANSWER. Write the word or phrase that best completes each statement or answers the question.

71) About 40% of the energy released in catabolism is used for cellular functions, and the rest is
_____.

Answer: given off as heat

72) The conversion of lactic acid to pyruvic acid is a(n) _____ reaction because it is a dehydrogenation reaction.

Answer: oxidation

73) In redox reactions, the _____ reaction is usually an exergonic reaction.

Answer: oxidation

74) The enzyme that is the key regulator of the rate of glycolysis is _____.

Answer: phosphofructokinase

75) The fate of pyruvic acid depends on the availability of _____.

Answer: oxygen

76) When acetylCoA enters the Krebs cycle, it combines with _____ to form _____.

Answer: oxaloacetic acid; citric acid

77) In the final step of the Krebs cycle, a hydrogen atom removed from malic acid is transferred to _____.

Answer: NAD^+

78) The most important outcome of the Krebs cycle are the reduced coenzymes _____ and _____.

Answer: NADH; $FADH_2$

79) The three proton pumps of the electron transport chain are the _____ complex, the _____ complex, and the _____ complex.

Answer: NADH dehydrogenase; cytochrome b–c; cytochrome oxidase

80) Glycogenesis is stimulated by the hormone _____.

Answer: insulin

81) Hepatocytes can release glucose into the blood, but skeletal muscles cannot because hepatocytes contain the enzyme _____ that converts glucose 6-phosphate to glucose.

Answer: phosphatase

82) Gluconeogenesis is stimulated by the hormones _____ and _____.

Answer: cortisol; glucagon

83) If _____ are present in excessive numbers, they deposit cholesterol in and around smooth muscle fibers in arteries to form plaques.

Answer: low-density lipoproteins.

84) The polar proteins making up the outer shell of a lipoprotein are called _____.

Answer: apoproteins

85) The apoprotein apo C-2 of chylomicrons and VLDLs activates the enzyme _____ , which triggers uptake of fatty acids by adipocytes.

Answer: endothelial lipoprotein lipase

86) A person is considered to have high blood cholesterol if total blood cholesterol is greater than _____ mg/dL.

Answer: 239

87) Rickets and osteomalacia are skeletal disorders resulting from deficiency of _____.

Answer: vitamin D

88) Macrocytic anemia and increased risk of neural tube defects are associated with deficiency of _____.

Answer: folic acid

89) A decrease in body temperature triggers a(n) _____ in the levels of thyroid hormones.

Answer: increase

90) Nonessential amino acids can be synthesized by the process of _____ , in which an amine group is added to pyruvic acid or other acids.

Answer: transamination

91) The rate at which the resting, fasting body breaks down nutrients to liberate energy is called the _____.

Answer: basal metabolic rate

92) Transfer of heat between objects without physical contact is called _____.

Answer: radiation

93) Formation of ATP by transfer of a high-energy phosphate group from an intermediate phosphorylated compound to ADP is called _____ phosphorylation.

Answer: substrate-level

94) Formation of ATP via energy released during the reactions of the electron transport chain is called _____ phosphorylation.

Answer: oxidative

95) Glycolysis is the oxidation of glucose to _____.

Answer: pyruvic acid

ESSAY. Write your answer in the space provided or on a separate sheet of paper.

96) Which vitamins are considered "antioxidant vitamins?" Why is this role so important?

Answer: Vitamins C, E, and beta-carotene (a provitamin) are antioxidants that inactivate oxygen free radicals. Free radicals damage cell membranes, DNA, and other cell structures. They also contribute to the formation of atherosclerotic plaque. Antioxidant vitamins may also decrease cancer risk, delay aging, and decrease the risk of cataract formation.

97) Describe the role of the hypothalamus in regulation of food intake.

Answer: The hypothalamus contains the neurons of the feeding center that stimulate eating and of the satiety center that signal fullness. It is thought that changes in blood chemistry (in terms of nutrients and hormone balance), as well as distention of the gastrointestinal tract, initiate appropriate hypothalamic activity.

98) Identify the different types of lipoproteins and describe the function of each.

Answer: Chylomicrons transport dietary lipids in the lymph and blood. Very low-density lipoproteins transport endogenous triglycerides from hepatocytes to adipocytes for storage. Low-density lipoproteins transport cholesterol through the body for use in repair of membranes and synthesis of steroid hormones and bile salts. High-density lipoproteins transport excess cholesterol to the liver for elimination.

99) Briefly outline the possible fates of glucose in the body.

Answer: 1) immediate oxidation for ATP production.
2) synthesis of amino acids for protein synthesis.
3) synthesis of glycogen for storage in liver and skeletal muscle.
4) formation of triglycerides via lipogenesis for long-term storage after glycogen stores are full.
5) excretion in urine if blood glucose is very high.

100) What are the possible fates of pyruvic acid in the body? What is the primary determinant of the fate of pyruvic acid? What is the fate of compounds to which pyruvic acid may be converted?

Answer: Pyruvic acid in the presence of low oxygen is reduced to lactic acid, which is converted to either glycogen or carbon dioxide. In the presence of high oxygen levels, pyruvic acid is converted to an acetyl unit, which may be carried into the Krebs cycle by coenzyme A or converted into fatty acids, ketone bodies, or cholesterol.

CHAPTER 26   The Urinary System

MULTIPLE CHOICE. Choose the one alternative that best completes the statements or answers the question.

1)   The renal corpuscle consists of
    A) the proximal and distal convoluted tubules.
    B) the glomerulus and the glomerular (Bowman's) capsule.
    C) the descending and ascending limbs of the loop of Henle.
    D) all the renal pyramids.
    E) the glomerulus and the vasa recta.

Answer: B

2)   Sympathetic nerves from the renal plexus are distributed to the
    A) renal blood vessels.
    B) convoluted tubules.
    C) renal pyramids.
    D) collecting ducts.
    E) both renal blood vessels and convoluted tubules.

Answer: A

3)   Which of the following lists the nephron regions in the correct order of fluid flow?
    A) glomerular capsule, distal convoluted tubule, loop of Henle, proximal convoluted tubule
    B) proximal convoluted tubule, loop of Henle, distal convoluted tubule, glomerular capsule
    C) glomerular capsule, proximal convoluted tubule, loop of Henle, distal convoluted tubule
    D) loop of Henle, glomerular capsule, proximal convoluted tubule, distal convoluted tubule
    E) distal convoluted tubule, loop of Henle, proximal convoluted tubule, glomerular capsule

Answer: C

4)   Which of the following lists the vessels in the correct order of blood flow?
    A) efferent arteriole, glomerulus, afferent arteriole, peritubular capillaries
    B) peritubular capillaries, efferent arteriole, glomerulus, afferent arteriole
    C) afferent arteriole, efferent arteriole, peritubular capillaries, glomerulus
    D) afferent arteriole, glomerulus, efferent arteriole, peritubular capillaries
    E) efferent arteriole, afferent arteriole, glomerulus, peritubular capillaries

Answer: D

5) The main function of the kidneys is to
   A) form urea from the metabolic breakdown of proteins.
   B) eliminate the excess water formed in metabolic processes.
   C) excrete extra-cellular fluid such as lymph.
   D) eliminate acid-forming carbon dioxide resulting from metabolism.
   E) regulate the composition of the blood and hence the whole internal environment.

   Answer: E

6) Bowman's capsule is
   A) the ball of capillaries from which the liquid part of the blood filters.
   B) another name for the kidney nephron.
   C) the connective tissue envelope surrounding the outside of the kidney.
   D) a double-walled funnel surrounding a glomerulus.
   E) All answers are correct.

   Answer: D

7) The unit excretory structure of the kidney is
   A) the renal pyramid.
   B) the glomerulus.
   C) the nephron.
   D) the neuron.
   E) the collecting duct.

   Answer: C

8) The part of a juxtamedullary nephron that is in the renal medulla is the
   A) glomerulus only.
   B) glomerular (Bowman's) capsule only.
   C) renal corpuscle.
   D) loop of Henle.
   E) entire nephron.

   Answer: D

9) The cells making up the proximal and distal convoluted tubules are
   A) stratified squamous epithelial cells.
   B) simple squamous epithelial cells.
   C) simple cuboidal epithelial cells.
   D) transitional epithelial cells.
   E) smooth muscle cells.

   Answer: C

10) Podocytes are cells specialized for filtration that are found in the
   A) walls of the vasa recta.
   B) ascending limb of the loop of Henle.
   C) urinary bladder.
   D) visceral layer of the glomerular capsule.
   E) collecting duct.

Answer: D

11) The surface of glomerular capillaries available for filtration is regulated by
   A) mesangial cells.
   B) macula densa cells.
   C) juxtaglomerular cells.
   D) renin.
   E) ADH.

Answer: A

12) ALL of the following are factors in the glomerular filter EXCEPT
   A) slit membranes.
   B) basal laminae.
   C) endothelial cells.
   D) fenestrations.
   E) microvilli.

Answer: E

13) If the diameter of the efferent arteriole is smaller than the diameter of the afferent arteriole, then
   A) blood pressure in the glomerulus stays low.
   B) blood pressure in the glomerulus stays high.
   C) there must be an abnormal blockage in the peritubular capillaries.
   D) the endothelial-capsular membrane filters less blood than normal.
   E) capsular hydrostatic pressure increases to levels higher than glomerular blood hydrostatic pressure.

Answer: B

14) Glomerular filtrate contains
   A) everything in blood.
   B) everything in blood except cells and proteins.
   C) water and electrolytes only.
   D) water and waste only.
   E) water only.

Answer: B

15) Filtration of blood in the glomeruli is promoted by
    A) blood colloid osmotic pressure.
    B) blood hydrostatic pressure.
    C) capsular hydrostatic pressure.
    D) both blood hydrostatic pressure and capsular hydrostatic pressure.
    E) both blood colloid osmotic pressure and capsular hydrostatic pressure.

Answer: B

16) Which of the following pressures is highest in the renal corpuscle under normal circumstances?
    A) blood colloid osmotic pressure
    B) capsular hydrostatic pressure
    C) capsular colloid osmotic pressure
    D) glomerular blood hydrostatic pressure
    E) None is higher than the others; all pressures are equal under normal circumstances.

Answer: D

17) An obstruction in the proximal convoluted tubule decreases glomerular filtration rate because
    A) blood hydrostatic pressure in the glomerulus decreases when blood can't flow through the tubule.
    B) osmotic pressure in the glomerular capsule increases due to leakage of more proteins into the filtrate.
    C) hydrostatic pressure in the glomerular capsule increases, which decreases net filtration pressure.
    D) hydrostatic pressure in the glomerular capsule decreases due to leakage of more filtrate into the peritubular space.
    E) release of renin decreases as fluid flow to the macula densa decreases.

Answer: C

18) The function of the macula densa cells is to
    A) prevent water reabsorption in the ascending limb of the loop of Henle.
    B) prevent over-distention of the urinary bladder.
    C) add bicarbonate ions to the tubular fluid in the proximal convoluted tubule.
    D) monitor NaCl concentration in the tubular fluid.
    E) produce the carrier molecules used to actively transport ions into the peritubular space.

Answer: D

19) In the myogenic mechanism of renal autoregulation
   A) renin causes contraction of macula densa cells to increase GFR.
   B) smooth muscle in afferent arterioles triggers vasoconstriction to decrease GFR.
   C) norepinephrine causes vasoconstriction of afferent arterioles to decrease GFR.
   D) atrial natriuretic peptide causes relaxation of mesangial cells to increase GFR.
   E) angiotensin II causes dilation of the proximal and distal convoluted tubules.

   Answer: B

20) The function of atrial natriuretic peptide in renal autoregulation of GFR is to stimulate
   A) renin secretion.
   B) conversion of angiotensin I to angiotensin II.
   C) relaxation of glomerular mesangial cells.
   D) constriction in afferent and efferent arterioles.
   E) reabsorption of sodium ions.

   Answer: C

21) If sympathetic stimulation to afferent and efferent arterioles decreases, then GFR
   A) doesn't change because the arterioles each have the same diameter.
   B) increases because the afferent arterioles dilate, but the efferent arterioles don't change.
   C) increases because both vessels are less constricted.
   D) decreases because both vessels constrict.
   E) doesn't change because the vessels do not have receptors for sympathetic neurotransmitters.

   Answer: C

22) The most important function of the juxtaglomerular (JG) apparatus is to
   A) secrete water and sodium into the tubular fluid.
   B) release renin in response to a drop in renal blood pressure or blood flow.
   C) make sure that the diameter of the efferent arteriole is kept larger than that of the afferent arteriole.
   D) produce antidiuretic hormone in response to increased glomerular filtration rate (GFR).
   E) produce chemicals that change the diameter of the loop of Henle.

   Answer: B

23) If there were an obstruction in the renal artery, one might expect to see:
   A) a decrease in glomerular filtration rate (GFR).
   B) an increase in the release of renin.
   C) an increase in glomerular filtration rate (GFR).
   D) Both B and C are correct.
   E) Both A and B are correct.

   Answer: E

24) As substances are reabsorbed in the proximal convoluted tubules of the kidneys, they move from
   A) filtered fluid to epithelial cells, to intersitial fluid to peritubular capillaries.
   B) filtered fluid to interstitial fluid, to epithelial cells, to peritubular capillaries.
   C) peritubular capillaries to interstitial fluid, to epithelial cells, to filtered fluid.
   D) vasa recta to epithelial cells, to interstitial fluid, to filtered fluid.
   E) peritubular capillaries to epithelial cells, to interstitial fluid, to filtered fluid.

   Answer: A

25) The uptake of substances from the lumen of the kidney tubules is known as
   A) tubular filtration.
   B) tubular secretion.
   C) tubular reabsorption.
   D) All answers are correct.

   Answer: C

26) Given the following conditions, glomerular blood hydrostatic pressure 75 mm Hg, capsular hydrostatic pressure 15 mm Hg, blood osmotic pressure 25 mm Hg, the effective filtration pressure would be _____mm Hg.
   A) 40
   B) 65
   C) 25
   D) 0
   E) 35

   Answer: E

27) Most reabsorption of substances from the glomerular filtrate occurs in the
   A) proximal tubule.
   B) loop of Henle.
   C) collecting tubule.
   D) distal tubule.
   E) Bowman's capsule.

   Answer: A

28) Which of the following substances is normally almost completely reabsorbed by the tubules of the nephron?
   A) creatinine
   B) glucose
   C) phosphate
   D) sodium
   E) urea

   Answer: B

29) The transport maximum is the
    A) highest the glomerular filtration rate can increase without inhibiting kidney function.
    B) greatest percentage of plasma entering the glomerulus that can become filtrate.
    C) upper limit of reabsorption due to saturation of carrier systems.
    D) steepest any concentration gradient can become.
    E) fastest rate at which fluid can flow through the renal tubules.

    Answer: C

30) Most water is reabsorbed in the proximal convoluted tubule by obligatory reabsorption, which means that
    A) water is moving up its own gradient.
    B) water is "following" sodium and other ions/molecules to maintain osmotic balance.
    C) the carrier that transports sodium cannot do so without binding water first.
    D) the proximal convoluted tubule cannot physically hold the volume of water that enters from the glomerular capsule, so water is reabsorbed because of hydrostatic pressure.
    E) the rate of water reabsorption never changes, regardless of water intake.

    Answer: B

31) Obligatory reabsorption of water occurs in the
    A) proximal convoluted tubule.
    B) distal convoluted tubule.
    C) ascending limb of the loop of Henle.
    D) descending limb of the loop of Henle.
    E) Both A and D are correct.

    Answer: E

32) Facultative reabsorption of water is regulated by
    A) angiotensin II.
    B) epinephrine.
    C) ADH.
    D) mesangial cells.
    E) calcitriol.

    Answer: C

33) Facultative reabsorption of water occurs mainly in the
    A) glomerulus.
    B) proximal convoluted tubule.
    C) descending limb of the loop of Henle.
    D) ascending limb of the loop of Henle.
    E) collecting ducts.

    Answer: E

34) Principal cells in the distal convoluted tubules
    A) secrete renin.
    B) monitor sodium and chloride ion concentrations in tubular fluid.
    C) secrete hydrogen ions when pH in the extracellular fluid is low.
    D) filter large proteins.
    E) respond to ADH and aldosterone.

    Answer: E

35) The significance of secretion of ammonium (NH_4^+) ions by the tubule cells is
    A) it triggers the release of renin.
    B) it results from generation of new bicarbonate ions that can be reabsorbed to help maintain pH.
    C) it keeps the ascending limb of Henle's loop from reabsorbing water.
    D) it carries urea across the endothelium of the vasa recta.
    E) there is no apparent function for this type of secretion.

    Answer: B

36) If the level of aldosterone in the blood increases, then
    A) more potassium is excreted in the urine.
    B) more sodium is excreted in the urine.
    C) blood pressure will drop.
    D) glomerular filtration rate will drop.
    E) First B, then C, then D.

    Answer: A

37) The amount of potassium secreted by principal cells is increased by which of the following?
    A) high levels of sodium ions in tubular fluid
    B) low levels of potassium in plasma
    C) the action of mesangial cells
    D) increased ADH
    E) Both A and D are correct.

    Answer: A

38) A role of intercalated cells is to
    A) secrete renin.
    B) secrete erythropoietin.
    C) reabsorb water in response to ADH.
    D) reabsorb sodium ions in response to aldosterone.
    E) excrete hydrogen ions when pH is too low.

    Answer: E

39) The action of ADH on principal cells is to
   A) increase production of sodium ion pumps.
   B) increase insertion of aquaporin-2 vesicles into apical membranes.
   C) increase the number of microvilli in their membranes.
   D) decrease the number of aquaporin-1 vesicles in basolateral membranes.
   E) do nothing because principal cells do not have ADH receptors.

   Answer: B

40) Urine that is hypotonic to blood plasma is produced when
   A) levels of aldosterone are high.
   B) levels of antidiuretic hormone are high.
   C) levels of antidiuretic hormone are low.
   D) plasma concentration of sodium ions is high.
   E) levels of both aldosterone and antidiuretic hormone are high.

   Answer: C

41) The concentration of solutes in tubular fluid is greatest in the
   A) glomerular (Bowman's) capsule.
   B) proximal convoluted tubule.
   C) hairpin turn of the loop of Henle.
   D) ascending limb of the loop of Henle.
   E) distal convoluted tubule.

   Answer: C

42) The renal clearance of a large protein such as albumin would be closest to which of the following values?
   A) the rate of renal blood flow
   B) the total blood volume entering both kidneys each minute
   C) the average glomerular filtration rate
   D) the transport maximum for glucose
   E) zero

   Answer: E

43) The permeability of the collecting ducts to water is regulated by
   A) aldosterone.
   B) renin.
   C) antidiuretic hormone.
   D) atrial natriuretic peptide.
   E) angiotensinogen.

   Answer: C

44) The countercurrent mechanism in the loop of Henle builds and maintains an osmotic gradient in the renal medulla. Which of the following is NOT a contributing factor?

A) Fluid flows in opposite directions in the ascending and descending limbs of the loop of Henle.

B) Chloride ions passively diffuse from the interstitial fluid into the thick portion of the ascending limb.

C) The thick portion of the ascending limb is impermeable to water.

D) The descending limb is permeable to water.

E) Fluid in the descending limb is in osmotic equilibrium with the surrounding interstitial fluid.

Answer: B

45) Cells that have receptors for aldosterone include

A) podocytes.

B) intercalated cells in the collecting ducts.

C) cells in the thick ascending limb of the loop of Henle.

D) cells in the proximal convoluted tubules.

E) cells in the distal convoluted tubules.

Answer: E

46) The effect of aldosterone on the principal cells of the distal convoluted tubule is to

A) increase the synthesis of sodium pumps.

B) increase the cells' permeability to water.

C) increase retention of potassium ions.

D) increase the cells' secretion of antidiuretic hormone.

E) trigger the release of renin.

Answer: A

47) The normal daily volume of urine produced is

A) under 200 ml.

B) 200–400 ml.

C) 1000–2000 ml.

D) 3 liters.

E) 180 liters.

Answer: C

48) Urea recycling in the renal medulla refers to the
   A) conversion of urea to ammonia by the tubule cells.
   B) conversion of ammonia to ammonium ions by the tubule cells.
   C) conversion of urea to amino acids in the vasa recta.
   D) mechanism by which urea leaves the collecting duct and re-enters the loop of Henle, thus helping to maintain the hypertonic conditions of the interstitial spaces.
   E) mechanism by which urea leaves the collecting ducts and enters the vasa recta, thus helping to maintain the correct blood volume in the vasa recta.

   Answer: D

49) Clearance refers to
   A) the volume of plasma from which a substance is entirely removed per minute.
   B) the amount of solute passed into the urine per minute.
   C) the amount of fluid passed across all the glomeruli per minute.
   D) the amount of solvent in the blood divided by the total volume.
   E) All answers are correct.

   Answer: A

50) Which of the following would be in the highest concentration in normal urine?
   A) albumin
   B) bilirubin
   C) creatinine
   D) acetoacetic acid
   E) urobilinogen

   Answer: C

MATCHING. Choose the item in column 2 that best matches each item in column 1.

51)	Tubular maximum.	A.	Juxtaglomerular apparatus
52)	Countercurrent multiplier system.	B.	Loop of Henle
53)	Aldosterone acts on	C.	Proximal convoluted tubule
54)	Renin is associated with	D.	Distal convoluted tubule
		E.	Collecting duct

Answers: 51) C. 52) B. 53) D. 54) A.

55)	Glomeruli are located in	A.	Cortex of the kidney
56)	Distal convoluted tubules are located in the	B.	Capsule of the kidney
57)	Loop of Henle is located in the	C.	Medulla of the kidney
58)	Ureter exits from the	D.	Pelvis of the kidney
		E.	Renal Hilus

Answers: 55) A. 56) A. 57) C. 58) E.

59) Juxtaglomerular apparatus produces

60) Associated with the hypothalamus.

61) Causes reabsorption of water.

62) ACE is associated with

A. Aldosterone

B. Parathyroid hormone

C. Antidiuretic hormone

D. Renin

E. Angiotensin II

Answers: 59) D. 60) C. 61) C. 62) E.

63) Used to measure glomerular filtration rate.

64) Catabolism of creatine phosphate from skeletal muscle.

65) Parathyroid hormones acts upon

66) Breakdown product of proteins.

A. Creatinine

B. Urea

C. Insulin

D. Calcium

E. All answers are correct

Answers: 63) C. 64) A. 65) D. 66) B.

67) Parathyroid hormone acts upon

68) Maintains an osmotic gradient.

69) Antidiuretic hormones acts upon

70) Acted upon by norepinephrine.

A. Glomerulus

B. Loop of Henle

C. Proximal convoluted tubules

D. Distal convoluted tubules

E. Collecting duct

Answers: 67) D. 68) B. 69) E. 70) A.

SHORT ANSWER. Write the word or phrase that best completes each statement or answers the questions.

71) The kidneys help regulate blood pressure by secretion of the enzyme _____ and by adjusting _____.

Answer: renin; renal resistance

72) The kidneys release two hormones: _____, which helps regulate calcium homeostasis, and _____, which increases red blood cell production.

Answer: calcitriol; erythropoietin

73) The functional units of the kidneys are the _____.

Answer: nephrons

74) Blood flows into afferent arterioles from _____.

Answer: interlobular arteries

75) The tubules of the juxtamedullary nephrons are served by special capillaries called _____.

Answer: vasa recta

76) Most renal nerves originate in the _____ ganglion, and their function is to regulate _____.

Answer: celiac; renal resistance and blood flow

77) Fluid flows from the ascending limb of the loop of Henle into the _____.

Answer: distal convoluted tubule

78) The juxtaglomerular apparatus consists of two parts: the _____ that detects the concentration of tubular fluid, and the _____ that secrete renin.

Answer: macula densa; juxtaglomerular cells

79) The last portion of the distal convoluted tubule and the collecting duct are made up of _____ cells, which are the target cells for ADH and aldosterone, and _____ cells that have microvilli and help regulate acid-base balance.

Answer: principal; intercalated

80) In tubular reabsorption, substances move from _____ to _____.

Answer: tubular fluid; blood

81) The footlike processes of podocytes are called _____, and the spaces between the processes are called _____.

Answer: pedicels; filtration slits

82) Glomerular endothelial cells are leaky because they have large pores called _____.

Answer: fenestrations

83) Norepinephrine causes _____ of afferent arterioles, which causes GFR to _____.

Answer: vasoconstriction; decrease

84) Angiotensin II causes _____ of the afferent arteriole and _____ of the efferent arteriole, which causes GFR to _____.

Answer: vasoconstriction; vasoconstriction; decrease

85) The hormone _____ increases capillary surface area available for filtration by causing relaxation of _____.

Answer: ANP; mesangial cells

86) If there is a decrease in the delivery of sodium and chloride ions to macula densa cells, tubuloglomerular feedback causes GFR to _____.

Answer: increase

87) The 10–12" tubes carrying urine from the kidneys to the urinary bladder are the _____.

Answer: ureters

88) The mucosa of the urinary bladder includes _____ epithelium.

Answer: transitional

89) The smooth muscle layers surrounding the mucosa of the urinary bladder are collectively known as the _____.

Answer: detrusor muscle

90) The normal component of urine that is derived from the detoxification of ammonia produced as a result of deamination of proteins is _____.

Answer: urea

91) The enzyme secreted by the juxtaglomerular cells in response to impulses from renal sympathetic nerves is _____.

Answer: renin

92) The substrate for the enzyme secreted by juxtaglomerular cells is _____.

Answer: angiotensinogen

93) The blood vessels surrounding the loop of Henle that help maintain the hypertonic conditions in the peritubular spaces of the renal medulla are called the _____.

Answer: vasa recta

94) The percentage of plasma in afferent arterioles that becomes glomerular filtrate is called the _____.

Answer: filtration fraction

95) In the formula for calculating net filtration pressure, those forces opposing glomerular filtration are _____ and _____.

Answer: capsular hydrostatic pressure; blood colloid osmotic pressure

ESSAY. Write your answer in the space provided or on a separate sheet of paper.

96) Describe the role of the kidney in acid-base balance.

Answer: The kidney can secrete hydrogen ions (via Na^+/H^+ antiporter). It can also reabsorb bicarbonate ions generated from carbon dioxide (from various sources) and generate new bicarbonate ions via deamination of glutamine.

97) Describe the flow of blood through the kidneys.

Answer: Kidneys receive 20–25% of the resting cardiac output via the renal arteries. The renal arteries branch to form segmental arteries, which branch to form interlobar arteries (through renal columns) to arcuate arteries (over bases of pyramids) to interlobular arteries. The interlobular arteries branch to form afferent arterioles to each nephron. Afferent arterioles branch to form glomerular capillaries where filtration occurs. Glomerular capillaries merge to form efferent arterioles, which then branch to form peritubular capillaries. Juxtamedullary nephrons also have vasa recta capillaries around them. Peritubular capillaries merge to form peritubular veins and with the vasa recta to form interlobular veins to arcuate veins to interlobar veins to segmental veins. Blood exits the kidney via renal veins.

98) Describe in detail the renin-angiotensin negative feedback loop that helps regulate blood pressure and glomerular filtration rate.

Answer: Stress causes a decrease in blood pressure, and thus, GFR. The JG cells of the juxtaglomerular apparatus sense decreased stretch and macula densa cells sense decreased NaCl and water. The JG cells secrete renin, which converts angiotensinogen in blood to angiotensin I, which is converted to angiotensin II by ACE in the lungs. Angiotensin II causes constriction of efferent arterioles, increased thirst, greater ADH secretion from the posterior pituitary, and increased secretion of aldosterone from the adrenal cortex. Blood volume is increased, which increases venous return, stroke volume, cardiac output, and blood pressure. GFR is also increased.

99) Discuss the importance of countercurrent flow to the functioning of the nephron.

Answer: Countercurrent flow refers to the flow of fluid in opposite directions in parallel tubing (tubules and blood vessels). The arrangement allows gradients to develop between tubular fluid, blood, and interstitial fluid. Gradients allow for reabsorption of large amounts of water and ions from the tubular fluid.

100) Describe the structural features of the renal corpuscle that enhance its blood-filtering capacity.

Answer: Endothelial cells of the glomerular capillaries are fenestrated. Their basement membranes are part of the filtering mechanism. Podocytes with filtration slits between pedicels wrap the glomerular capillaries. The large surface area also contributes to filtering ability, as does the high glomerular hydrostatic pressure created by the arrangement of the afferent and efferent arterioles, in which the diameter of the efferent arteriole is smaller than that of the afferent arteriole.

CHAPTER 27   Fluid, Electrolyte, and Acid-Base Homeostasis

MULTIPLE CHOICE. Choose the one alternative that best completes the statement or answers the question.

1) The primary means of water movement between fluid compartments is
    A) osmosis.
    B) primary active transport.
    C) secondary active transport.
    D) facilitated diffusion.
    E) pinocytosis.
    Answer: A

2) The direction of water movement between fluid compartments is determined by
    A) the electrical gradient.
    B) the solubility of water in membrane lipids.
    C) the concentration of solutes.
    D) the diameter of blood vessels.
    E) differences in pH.
    Answer: C

3) Women generally have a lower amount of total body water than men because
    A) they are smaller than men.
    B) estrogen causes greater water loss than testosterone.
    C) they have a higher body temperature.
    D) they have a higher percentage of body fat.
    E) All of these are correct.
    Answer: C

4) The thirst center is stimulated by ALL of the following EXCEPT
    A) osmoreceptors in the hypothalamus.
    B) peripheral chemoreceptors.
    C) baroreceptors.
    D) dry mouth.
    E) angiotensin II.
    Answer: B

5) The primary determinant of body fluid volume is the
   A) concentration of potassium ions inside cells.
   B) level of physical activity.
   C) amount of water ingested.
   D) body weight.
   E) number of sodium and chloride ions lost from the kidney.

   Answer: E

6) The stimulus for release of ANP is
   A) renin.
   B) ADH.
   C) aldosterone.
   D) stretching of the atrial wall.
   E) osmoreceptors.

   Answer: D

7) A decrease in angiotensin II leads to
   A) increased blood volume due to decreased GFR.
   B) decreased blood volume due to increased GFR.
   C) decreased blood volume due to decreased GFR.
   D) increased blood volume due to increased GFR.
   E) no changes in blood volume.

   Answer: B

8) The area that stimulates the conscious desire to drink water is located in the
   A) adrenal cortex.
   B) kidney.
   C) medulla oblongata.
   D) hypothalamus.
   E) lumbar region of the spinal cord.

   Answer: D

9) In studies of fluid balance, the term *water intoxication* refers to
   A) poisoning of the body's water due to buildup of toxic substances during renal failure.
   B) increased blood hydrostatic pressure created by high total blood volume.
   C) movement of water from interstitial fluid into intracellular fluid due to osmotic gradients created by ion loss.
   D) any situation in which edema develops.
   E) failure of the neurohypophysis to secrete sufficient ADH.

   Answer: C

10) ADH saves water by
   A) promoting the excretion of sodium ions.
   B) stimulation the secretion of renin.
   C) enhancing passive movement of water out of the collecting ducts.
   D) stimulating constriction of the lumen of the distal convoluted tubules.
   E) lowering the glomerular filtration rate.

   Answer: C

11) Extracellular fluids are
   A) high in both sodium and potassium.
   B) low in both sodium and potassium.
   C) high in sodium and low in potassium.
   D) low in sodium and high in potassium.

   Answer: C

12) Excessive intake or drinking of water normally leads to
   A) hypertonicity of the blood.
   B) increased permeability of the collecting duct to water.
   C) decreased blood volume.
   D) reduced ADH secretion.
   E) All of the answers are correct.

   Answer: D

13) The cation that is necessary for generation and conduction of action potentials and that contributes nearly half of the osmotic pressure of extracellular fluid is
   A) sodium ion.
   B) potassium ion.
   C) calcium ion.
   D) chloride ion.
   E) phosphate ion.

   Answer: A

14) Levels of sodium ions in the extracellular fluid are regulated primarily by
   A) ADH.
   B) aldosterone.
   C) parathyroid hormone.
   D) epinephrine.
   E) insulin.

   Answer: B

15) Levels of potassium ions in the extracellular fluid are regulated primarily by
   A) ADH.
   B) aldosterone.
   C) parathyroid hormone.
   D) epinephrine.
   E) insulin.

   Answer: B

16) The primary intracellular ions are
   A) potassium and chloride ions and protein anions.
   B) sodium and phosphate ions.
   C) potassium and phosphate ions and protein anions.
   D) sodium and chloride ions.
   E) potassium, phosphate, and calcium ions.

   Answer: C

17) Drinking plain water after excessive sweating leads to
   A) shut-down of sweat glands.
   B) hypernatremia.
   C) water intoxication.
   D) dehydration of cells.
   E) Both C and D are correct.

   Answer: C

18) Protein anions are most abundant in
   A) plasma.
   B) interstitial fluid.
   C) the cytosol.
   D) urine.
   E) glomerular filtrate.

   Answer: C

19) When bicarbonate ion diffuses out of red blood cells into plasma, it is usually exchanged with which anion?
   A) sodium
   B) potassium
   C) phosphate
   D) hydrogen
   E) chloride

   Answer: E

20) Hypernatremia can be defined as a(n)

   A) reduction of concentration of plasma calcium.

   B) decrease in plasma potassium.

   C) increase in plasma phosphate.

   D) excessive retention of plasma sodium.

   E) All of the answers are correct.

   Answer: D

21) Which of the following would you expect to see in response to an extracellular fluid calcium ion level of 5.7 mEq/liter?

   A) increased secretion of aldosterone

   B) increased secretion of PTH

   C) increased secretion of CT

   D) decreased secretion of ANP

   E) increased secretion of ADH

   Answer: C

22) Bicarbonate ion acts as a

   A) nonelectrolyte.

   B) strong acid.

   C) strong base.

   D) weak acid.

   E) weak base.

   Answer: E

23) The carboxyl group of an amino acid acts as a buffer for

   A) excess hydrogen ions.

   B) excess hydroxide ions.

   C) other carboxyl groups.

   D) carbonic acid in red blood cells.

   E) hydrochloric acid in gastric juice.

   Answer: B

24) Why are levels of bicarbonate ion higher in arterial blood than in venous blood?

   A) because the partial pressure of carbon dioxide is higher in arterial blood

   B) because more bicarbonate ions are used up in venous blood to buffer hydrogen ions

   C) because the higher oxygen levels in arterial blood promote dissociation of carbonic acid

   D) because cells in the pulmonary capillaries actively secrete bicarbonate ions into the plasma

   E) because the higher oxygen levels in arterial blood increase the activity of carbonic anhydrase

   Answer: B

25) Which of the following is NOT an effect of increased levels of parathyroid hormone?
   A) increased absorption of calcium ions from the gastrointestinal tract
   B) increased reabsorption of calcium ions by renal tubule cells
   C) increased reabsorption of phosphate ions by renal tubule cells
   D) increased release of calcium ions from mineral salts in bone matrix
   E) increased release of phosphate ions from mineral salts in bone matrix
   Answer: C

26) Hemoglobin picks up a hydrogen ion when
   A) it releases oxygen to tissues.
   B) it binds oxygen in pulmonary capillaries.
   C) chloride ions enter red blood cells.
   D) the intracellular concentration of monohydrogen phosphate ions is too low to be effective.
   E) chloride ions leave red blood cells.
   Answer: A

27) Which of the following statements is correct?
   A) A strong acid plus a weak acid yields water plus a weak base.
   B) A strong acid plus a weak base yields a salt plus a weak acid.
   C) A strong acid plus a weak base yields a weak base plus a weak acid.
   D) A strong acid plus a strong base yields a weak acid plus a weak base.
   E) A strong acid plus a weak acid yields a strong base plus a weak base.
   Answer: B

28) The ratio of bicarbonate ions to carbonic acid molecules in extracellular fluid is normally about
   A) 1:20.
   B) 1:1.
   C) 2:1.
   D) 20:1.
   E) 100:1.
   Answer: D

29) Which of the following cannot help protect against pH changes caused by respiratory problems in which there is an excess or shortage of carbon dioxide?
   A) plasma protein buffers
   B) hemoglobin
   C) bicarbonate ion/carbonic acid buffers
   D) phosphate buffers
   E) Only phosphate buffers can help protect against such pH changes.
   Answer: C

30) An acid may be defined as
   A) a donor of hydrogen.
   B) an acceptor of hydrogen.
   C) a material that is completely dissociated.
   D) a material that is not completely dissociated.

   Answer: A

31) If the blood is acidic, which one of the following would NOT occur?
   A) Ammonia would be secreted by the cells of the kidney tubules.
   B) The person's rate of breathing would increase.
   C) Sodium ions would be taken up by the kidneys.
   D) Hydrogen ions would be excreted by the kidneys.
   E) All of the answers are correct.

   Answer: E

32) A pH of 6.5 is said to be
   A) alkaline.
   B) acid.
   C) neutral.

   Answer: B

33) Holding your breath for an extended period of time results in
   A) respiratory acidosis.
   B) respiratory alkalosis.
   C) metabolic acidosis.
   D) metabolic alkalosis.

   Answer: A

34) Hyperventilation results in
   A) metabolic acidosis.
   B) respiratory acidosis.
   C) metabolic alkalosis.
   D) respiratory alkalosis.

   Answer: D

35) In compensating for respiratory alkalosis, the body excretes more
   A) ammonium ions.
   B) bicarbonate ions.
   C) dihydrogen phosphate ions.
   D) carbonic acid.
   E) hydrogen ions.

   Answer: B

36) An increase in ADH leads to
   A) insertion of aquaporin-2 channels into principal cell membranes.
   B) an increase in aldosterone.
   C) stimulation of the thirst center.
   D) excretion of bicarbonate ions.
   E) Both A and C are correct.

   Answer: A

37) Which of the following statements is correct?
   A) A strong base plus a weak base yields a salt plus a weak base.
   B) A strong base plus a weak acid yields a strong acid and a weak base.
   C) A strong base plus a strong acid yields a weak base plus a weak acid.
   D) A strong base plus a weak acid yields water plus a weak base.
   E) A strong base plus a weak base yields a strong acid plus a weak acid.

   Answer: D

38) The inspiratory center in the medulla oblongata triggers more forceful and frequent contractions of the diaphragm if
   A) a decrease in pCO_2 is detected by peripheral chemoreceptors.
   B) a large quantity of an alkaline drug is ingested.
   C) levels of ketone bodies become elevated.
   D) hydrochloric acid is lost via severe vomiting.
   E) blood pressure is increased.

   Answer: C

39) Hydrogen ions are normally eliminated from the body
   A) by excretion in urine.
   B) via insensible perspiration.
   C) in expired air.
   D) Both A and B are correct.
   E) Both A and C are correct.

   Answer: A

40) A person who has not eaten for a week is probably
   A) generating ketone bodies.
   B) excreting excess hydrogen ions.
   C) generating new bicarbonate ions.
   D) breathing more rapidly than normal.
   E) All of these are correct.

   Answer: E

41) In compensating for metabolic acidosis, the body
   A) increases respiratory rate.
   B) excretes more bicarbonate ions.
   C) excretes more monohydrogen phosphate ions.
   D) decreases respiratory rate.
   E) slows the rate of conversion of ammonia to urea.

   Answer: A

42) If the pH of blood plasma becomes 7.49 due to ingested substances, ALL of the following would happen to compensate EXCEPT
   A) respiratory rate decreases.
   B) the kidney increases excretion of bicarbonate ions.
   C) tubule cells produce more ammonia from glutamate.
   D) the partial pressure of carbon dioxide in blood would begin to rise.
   E) the kidney excretes fewer dihydrogen phosphate ions.

   Answer: C

43) Aldosterone regulates the level of chloride ions in body fluids by
   A) opening chloride channels in principal cells of distal convoluted tubules.
   B) reabsorbing chloride ions for electrical balance as bicarbonate ions are secreted from renal tubules.
   C) altering the permeability of glomerular capillaries.
   D) regulating secretion from gastric mucosal glands.
   E) controlling reabsorption of sodium ions, which chloride ions follow due to electrical attraction.

   Answer: E

44) Increasing respiratory rate will
   A) add more hydrogen ions to the extracellular fluid.
   B) result in an increase in excretion of excess bicarbonate ions in urine.
   C) result in an increase in excretion of dihydrogen phosphate ions in urine.
   D) lower the pH of extracellular fluid.
   E) cause a decrease in the affinity of hemoglobin for oxygen.

   Answer: B

45) Uncontrolled diabetes mellitus may lead to metabolic acidosis because
   A) high glucose levels depress the respiratory centers in the medulla.
   B) glucose is an acidic substance.
   C) glucose is osmotically active, and for every water molecule retained, a hydrogen ion is also retained.
   D) most diabetics have chronic diarrhea, which leads to excessive loss of bicarbonate ions.
   E) increased rates of lipolysis and ketogenesis occur.

   Answer: E

46) Which of the following might trigger an increase in the rate of deamination of glutamine by renal tubule cells as a form of compensation for a pH imbalance?
   A) an abrupt move to a high altitude
   B) ingestion of alkaline drugs
   C) plasma levels of bicarbonate ion at 30 mEq/liter
   D) pulmonary edema
   E) severe, prolonged vomiting

   Answer: D

47) A patient whose blood pH is 7.47, whose pCO_2 is 31 mmHg in arterial blood, and whose levels of bicarbonate ion in arterial blood are 23 mEq/liter is in
   A) compensated metabolic alkalosis.
   B) uncompensated respiratory acidosis.
   C) uncompensated respiratory alkalosis.
   D) uncompensated metabolic acidosis.
   E) uncompensated metabolic alkalosis.

   Answer: C

48) The most abundant buffer intracellular fluid system is
    A) proteins.
    B) phosphates.
    C) bicarbonates.
    D) ammonium.
    E) All answers are equal.
    Answer: A

49) Use of laxatives in older people to relieve constipation often results in
    A) hyponatremia.
    B) hypokalemia.
    C) hypernatremia.
    D) hypocalcemia.
    Answer: B

50) Edema may result from
    A) hypoproteinemia.
    B) lymphatic blockage.
    C) increased blood hydrostatic pressure.
    D) All of the above are correct.
    Answer: D

MATCHING. Choose the item in column 2 that best matches each item in column 1.

51) Excess sodium.   A. Hypocalcemia
52) Depletion of sodium.   B. Hyponatremia
53) Calcium depletion.   C. Hypernatremia
54) Excess potassium.   D. Hyperkalemia
                                     E. Hypokalemia

    Answers: 51) C. 52) B. 53) A. 54) D.

55) Water control in collecting duct.   A. Aldosterone
56) Controls calcium.   B. Parathyroid hormone
57) Controls sodium.   C. Insulin
58) Controls glucose.   D. Antidiuretic hormone
                                         E. Glucocorticoid

    Answers: 55) D. 56) B. 57) A. 58) C.

59) Excessive vomiting with substantial loss of HCL.

A. Metabolic acidosis

60) Accumulation of acids like ketosis.

B. Respiratory alkalosis

61) Hyperventilation.

C. Metabolic alkalosis

62) Results from emphysema.

D. Respiratory acidosis

Answers: 59) C. 60) A. 61) B. 62) D.

63) Highest cation intracellularly.

A. Sodium

64) Highest cation extracellularly.

B. Potassium

65) Highest cation in plasma.

C. Chloride

66) Highest anion extracellularly.

D. Calcium

E. Sulfate

Answers: 63) B. 64) A. 65) A. 66) C.

67) Most common cation in bone.

A. Phosphate

68) Most common anion in bone.

B. Chloride

69) Cofactor for sodium pump.

C. Calcium

70) Exchanged across the membrane for bicarbonate.

D. Magnesium

E. Sodium

Answers: 67) C. 68) A. 69) D. 70) B.

SHORT ANSWER. Write the word or phrase that best completes each statement or answers the question.

71) Severe vomiting leads to the pH imbalance _____ due to loss of _____.

Answer: metabolic alkalosis; gastric acids

72) Aspirin overdose leads to the pH imbalance _____ due to _____.

Answer: respiratory alkalosis; hyperventilation (decreased partial pressure of carbon dioxide)

73) You would expect a person with chronic obstructive pulmonary disease to be excreting more _____ in urine than a healthy person.

Answer: hydrogen ions

74) The physiological response to an acid-base imbalance that acts to normalize arterial blood pH is called _____.

Answer: compensation

75) In the phosphate buffer system, the weak acid is _____ and the weak base is _____.

Answer: dihydrogen phosphate; monohydrogen phosphate

76) The most abundant buffer in intracellular fluid and plasma is the _____ system.

Answer: protein buffer

77) The _____ group of an amino acid acts as an acid.

Answer: carboxyl

78) Hydrogen ions are buffered inside red blood cells by _____.

Answer: reduced hemoglobin

79) Bone resorption is stimulated by the hormone _____ to raise the blood levels of _____ ions.

Answer: parathyroid hormone; calcium

80) The normal range of plasma concentration of bicarbonate ions is _____ in arterial blood; it is slightly _____ in venous blood.

Answer: 22–26 mEq/liter; higher

81) PTH stimulates _____ of calcium ions from kidney tubules, _____ of phosphate ions, and _____ of magnesium ions.

Answer: reabsorption; excretion; reabsorption

82) The level of sodium ions in the blood is controlled by the hormones _____, _____, and _____.

Answer: aldosterone; ANP; ADH

83) The condition resulting when water intake exceeds the kidneys' excretory ability is called _____.

Answer: water intoxication

84) The thirst center is stimulated by _____, _____, and _____.

Answer: dry mouth; angiotensin II; osmoreceptors in the hypothalamus

85) In a healthy, resting person the greatest loss of water other than urinary loss is by _____.

Answer: evaporation from the skin

86) The major extracellular fluid compartments are the _____ and _____.

Answer: interstitial fluid; plasma

87) Plasma levels of potassium ions are regulated primarily by the hormone _____.

Answer: aldosterone

88) As age increases, the percentage of body weight that is water _____; as the amount of adipose tissue increases, the percentage of body weight that is water _____.

Answer: decreases; decreases

89) The positive or negative charge equal to the amount of charge in one mole of hydrogen ions is called one _____.

Answer: equivalent

90) In intracellular fluid, the most abundant cation is _____ and the most abundant inorganic anion is _____.

Answer: potassium; monohydrogen phosphate

91) The most abundant extracellular anion is _____.

Answer: chloride ion

92) As blood passes through the pulmonary capillaries, the plasma level of bicarbonate ion _____.

Answer: decreases

93) The homeostatic range of pH for extracellular fluid is _____ to _____.

Answer: 7.35; 7.45

94) When respiratory rate increases, pH of extracellular fluid _____.

Answer: increases

95) A strong acid combined with a weak base yields _____ and _____.

Answer: water; a weak acid

ESSAY. Write your answer in the space provided or on a separate sheet of paper.

96) Describe the fluid and electrolyte disorders to which the elderly are particularly susceptible.

Answer: 1) Dehydration and hypernatremia due to inadequate fluid intake or loss of more water than sodium in vomit, feces, or urine.

2) Hyponatremia due to inadequate intake of sodium, impaired kidney function, or excessive sodium loss.

3) Hypokalemia due to excessive laxative use or potassium-depleting diuretics.

4) Acidosis due to lung or kidney disease.

97) Explain how it is possible for a patient with chronic obstructive pulmonary disease to have a normal extracellular pH while having an elevated partial pressure of carbon dioxide.

Answer: Elevated partial pressure of carbon dioxide causes respiratory acidosis, which is compensated by an increase in plasma levels of bicarbonate ion. Because the patient cannot breathe off the excess carbon dioxide due to structural changes in the respiratory system, the partial pressure of carbon dioxide stays high, but compensated by bicarbonate.

98) A patient's blood pH is 7.48; partial pressure of carbon dioxide is 32 mm Hg and levels of bicarbonate in the blood are 20 mEq/liter. What can you tell about this patient's condition? Explain your answer.

Answer: The patient is in respiratory alkalosis (high pH, low carbon dioxide), which is partially compensated (low bicarbonate).

99) Describe the negative feedback loop that stimulates thirst as a result of dehydration.

Answer: Dehydration causes 1) decreased flow of saliva, which dries the mouth and pharynx, 2) increased blood osmotic pressure, which stimulates osmoreceptors in the hypothalamus, and 3) decreased blood volume, which lowers blood pressure, increasing release of renin from JG cells, increasing levels of angiotensin II. All of these stimulate the thirst center in the hypothalamus, which increases fluid intake via thirst, thus increasing body water.

100) What are acidosis and alkalosis, and how do they develop? What are the primary effects of each?

Answer: Acidosis is blood pH under 7.35, which causes depression of the CNS via reduced synaptic transmission. Alkalosis is blood pH over 7.45, which causes overexcitability of the nervous system. Anything that causes partial pressure of carbon dioxide to rise or causes loss of bicarbonate ions or causes buildup of metabolic acids leads to acidosis. Anything that lowers partial pressure of carbon dioxide or causes loss of acids or any ingestion of alkaline substances leads to alkalosis.

CHAPTER 28   The Reproductive Systems

MULTIPLE CHOICE. Choose the one alternative that best completes the statement or answers the question.

1)   The function of the cremaster muscle is to
   A) elevate the testes during sexual arousal and exposure to cold.
   B) generate peristaltic waves in the ductus deferens.
   C) control the release of secretions from the seminal vesicles.
   D) control the release of sperm cells from the testes into the epididymis.
   E) prevent urine from entering the urethra during ejaculation.

   Answer: A

2)   Sertoli cells produce
   A) testosterone.
   B) androgen-binding protein.
   C) estrogen.
   D) FSH.
   E) LH.

   Answer: B

3)   Testosterone is produced by
   A) spermatozoa.
   B) sustentacular cells.
   C) interstitial cells.
   D) the hypothalamus.
   E) all cells in the male.

   Answer: C

4)   Leydig cells are located
   A) in all the male accessory reproductive organs.
   B) interspersed among developing sperm cells in seminiferous tubules.
   C) lining the epididymis and ductus deferens.
   D) within the tunica albuginea.
   E) in spaces between adjacent seminiferous tubules.

   Answer: E

5) Sperm production in the male requires a scrotal temperature that is
   A) higher than body temperature.
   B) the same as body temperature.
   C) lower than body temperature.
   D) independent of body temperature.
   E) variable between day and night.

   Answer: C

6) The immune system does not normally attack spermatogenic cells because
   A) they are recognized as "self" structures.
   B) they do not have any antigens on their cell membranes.
   C) spermatogenic cells are protected by the blood-testis barrier.
   D) the acrosome covers any antigens that would be recognized as foreign.
   E) spermatogenic cells are release chemicals that repel antigen-presenting cells.

   Answer: C

7) The cells that result from the equatorial division of spermatogenesis are called
   A) spermatogonia.
   B) primary spermatocytes.
   C) secondary spermatocytes.
   D) primordial germ cells.
   E) spermatids.

   Answer: E

8) The process of spermiation is the
   A) reduction division of male gamete production.
   B) equatorial division of male gamete production.
   C) process producing the liquid portion of semen.
   D) release of a sperm cell from its connection to a sustentacular cell.
   E) change that occurs in the sperm cell following penetration of the egg.

   Answer: D

9) The process of crossing-over, or recombination, of genes occurs during
   A) meiosis I.
   B) meiosis II.
   C) spermiogenesis.
   D) spermiation.
   E) fertilization.

   Answer: A

10) The form (stag) of developing male gamete located nearest to the basement membrane of a seminiferous tubule is the
A) spermatid.
B) primary spermatocyte.
C) secondary spermatocyte.
D) primordial germ cell.
E) spermatogonium.

Answer: E

11) During spermatogenesis, which of the following undergoes a meiotic division to produce haploid cells?
A) spermatids
B) secondary spermatocytes
C) primary spermatocytes
D) spermatogonia
E) spermatozoa

Answer: C

12) Final maturation of sperm cells occurs in the
A) epididymis.
B) seminiferous tubules.
C) prostate gland.
D) urethra.
E) female reproductive tract.

Answer: A

13) Which of the following cells are diploid?
A) secondary oocytes
B) secondary spermatocytes
C) primary spermatocytes
D) spermatids
E) All of the above except spermatids.

Answer: C

14) The acrosome of a sperm cell contains
A) the chromosomes.
B) mitochondria for energy production.
C) testosterone.
D) hyaluronidase for egg penetration.
E) the flagellum.

Answer: D

15) Which of the following cells are diploid?
   A) secondary oocytes
   B) secondary spermatocytes
   C) primary spermatocytes
   D) spermatids
   E) All of the above except spermatids.

   Answer: C

16) A function of FSH in the male is to
   A) inhibit progesterone.
   B) initiate testosterone production.
   C) increase protein synthesis.
   D) inhibit estrogen.
   E) initiate spermatogenesis.

   Answer: E

17) In the male, LH causes
   A) initiation of spermatogenesis.
   B) development of secondary sex characteristics.
   C) testosterone production.
   D) ejaculation.
   E) release of GnRH.

   Answer: C

18) The principal androgen is
   A) ABP.
   B) FSH.
   C) testosterone.
   D) hCG.
   E) estradiol.

   Answer: C

19) Seminal vesicles produce
   A) sperm cells.
   B) testosterone.
   C) fructose-rich fluid.
   D) estrogen.
   E) mucus.

   Answer: C

20) The normal number of spermatozoa per milliliter of semen is
   A) 50–100.
   B) fewer than 20,000,000.
   C) more than 200,000,000.
   D) 50,000,000–150,000,000.
   E) about 5 million.

   Answer: D

21) The function of fructose in semen is to
   A) provide an energy source for ATP production by sperm.
   B) promote coagulation of semen in the female reproductive tract.
   C) buffer acids in the female reproductive tract.
   D) inhibit the growth of bacteria in semen and the female reproductive tract.
   E) provide an energy source for the zygote.

   Answer: A

22) The seminal vesicles are located
   A) inferior to the prostate within the urogenital diaphragm.
   B) within the lobules of the testes.
   C) within the spermatic cord.
   D) posterior and inferior to the urinary bladder, in front of the rectum.
   E) on the posterior surface of each testis.

   Answer: D

23) Which of the following does NOT manufacture products that become part of semen?
   A) seminiferous tubules
   B) bulbourethral glands
   C) penis
   D) seminal vesicles
   E) prostate gland

   Answer: C

24) A normal mature human spermatozoa contains
   A) 23 chromosomes.
   B) 23 pairs of chromosomes.
   C) 46 chromosomes.
   D) 46 pairs of chromosomes.
   E) All of the above are possible.

   Answer: A

25) Interstitial cells (of Leydig)

A) produce sperm.

B) secrete testosterone.

C) dissolve sperm that are not ejaculated.

D) produce ova.

E) secrete estrogen

Answer: B

26) The female structure that is homologous to the penis is the

A) ovary.

B) uterus.

C) vagina.

D) clitoris.

E) Bartholin's gland.

Answer: D

27) The glycoprotein layer between the oocyte and the granulosa cells of an ovarian follicle is called the

A) theca interna.

B) theca externa.

C) antrum.

D) zona pellucida.

E) corona radiata.

Answer: D

28) The oogenesis begins in females

A) before birth.

B) only if the egg is fertilized.

C) after ovulation.

D) monthly after puberty in response to FSH and LH.

E) when adrenal gonadocorticoids begin to rise at the start of puberty.

Answer: A

29) The secretory cells of an ovarian follicle are called the

A) theca interna.

B) theca externa.

C) antrum.

D) zona pellucida.

E) corona radiata.

Answer: A

30) Which of the following help move the oocyte into and through the uterine tube?
   A) peristalsis
   B) cilia
   C) flagella
   D) fimbriae
   E) All of the above except flagella.

Answer: E

31) The opening between the cervical canal and the uterine cavity is called the
   A) internal os.
   B) external os.
   C) isthmus.
   D) fornix.
   E) vagina.

Answer: A

32) The folds of the peritoneum attaching the uterus to either side of the pelvic cavity are called the
   A) uterosacral ligaments.
   B) broad ligaments.
   C) cardinal ligaments.
   D) round ligaments.
   E) suspensory ligaments.

Answer: B

33) Which of the following lists the uterine blood vessels in the correct order of blood flow?
   A) radial arteries, arcuate arteries, spiral arterioles, straight arterioles
   B) spiral arterioles, straight arterioles, radial arteries, arcuate arteries
   C) straight arterioles, radial arteries, arcuate arteries, spiral arterioles
   D) arcuate arteries, radial arteries, straight arterioles, spiral arterioles
   E) arcuate arteries, straight arterioles, spiral arterioles, radial arteries

Answer: D

34) The epithelium of the vaginal mucosa is
   A) simple squamous.
   B) simple cuboidal.
   C) simple columnar.
   D) transitional.
   E) stratified squamous.

Answer: E

35) The female structure that is homologous to the scrotum is the

A) mons pubis.

B) labia majora.

C) labia minora.

D) clitoris.

E) hymen.

Answer: B

36) The perineum is bounded by the

A) pubic symphysis, iliac crests, and sacral promontory.

B) anterior and posterior inferior iliac spines, and pubic symphysis.

C) pubic symphysis, ischial tuberosities, and coccyx.

D) pubic symphysis, posterior inferior iliac spines, and coccyx.

E) ischial tuberosities, iliac crests, and anterior inferior iliac spines.

Answer: C

37) ALL of the following are functions of estrogens EXCEPT

A) help control fluid and electrolyte balance.

B) promote protein anabolism.

C) help regulate secretion of FSH.

D) promote development and maintenance of female secondary sex characteristics.

E) raise blood cholesteol.

Answer: E

38) The part of the female reproductive system that is shed during menstruation is the

A) myometrium.

B) mucosa of the vagina.

C) tunica albuginea.

D) stratum functionalis of the endometrium.

E) germinal epithelium.

Answer: D

39) During the menstrual cycle, the endometrium would be at its thickest

A) during the menstrual phase.

B) just prior to ovulation.

C) just after ovulation.

D) late in the postovulatory phase.

E) The thickness never changes.

Answer: D

40) During the menstrual cycle, LH is at its highest levels
   A) during the menstrual phase.
   B) just prior to ovulation.
   C) just after ovulation.
   D) just before menstruation begins.
   E) Levels of LH never change.

   Answer: B

41) During the menstrual cycle, progesterone would be at its highest levels
   A) during the menstrual phase.
   B) just prior to ovulation.
   C) just after ovulation.
   D) late in the postovulatory phase.
   E) Levels of progesterone never change.

   Answer: D

42) Repair of the endometrium during the preovulatory phase of menstruation is due to rising levels of
   A) FSH.
   B) estrogen.
   C) hCG.
   D) progesterone.
   E) inhibin.

   Answer: B

43) During the menstrual cycle, progesterone is produced by
   A) the secondary oocyte.
   B) the corpus luteum.
   C) the stroma of the ovary.
   D) primary follicles.
   E) the endometrium

   Answer: B

44) The main function of progesterone during the menstrual cycle is to
   A) initiate ovulation.
   B) initiate menstruation.
   C) thicken the endometrium.
   D) repair the surface of the ovary after ovulation.
   E) stimulate the release of FSH and LH.

   Answer: C

45) If fertilization does not occur, the corpus luteum:
    A) is expelled into the pelvic cavity.
    B) begins to secrete low levels of FSH.
    C) degenerates into the corpus albicans.
    D) continues to secrete progesterone until the next ovulation.
    E) Both A and C are correct.

    Answer: C

46) ALL of the following are sympathetic responses during sexual intercourse EXCEPT
    A) peristalsis in the ductus deferens.
    B) increased blood pressure.
    C) contraction of perineal muscles.
    D) ejaculation of semen.
    E) erection of the penis/clitoris.

    Answer: E

47) Oral contraceptives for women typically contain
    A) human chorionic gonadotropin.
    B) progestin and estrogen.
    C) low levels of both FSH and LH.
    D) nonoxynol-9.
    E) testosterone.

    Answer: B

48) Onset of puberty in both sexes is signaled by increases in levels of
    A) estradiol.
    B) DHT.
    C) oxytocin.
    D) PRL.
    E) LH.

    Answer: E

49) The main control center of the female reproductive system is the
    A) anterior pituitary.
    B) posterior pituitary.
    C) ovary.
    D) hypothalamus.
    E) penis.

    Answer: D

50) The corpus luteum produces
    A) testosterone.
    B) menstrual fluid.
    C) progesterone.
    D) mucus.
    E) All of the answers are correct.

    Answer: C

MATCHING. Choose the item in column 2 that best matches each item in column 1.

51) External sac enclosing the testis.   A. Uterus
52) Place where the fetus develops and grows.   B. Labia majora
53) Female homologue of the scrotum.   C. Scrotum
54) Part of the body of the penis.   D. Corpus cavernosa
                                                E. Clitoris

    Answers: 51) C. 52) A. 53) B. 54) D.

55) Stimulates spermatogenesis.   A. hCG
56) Secreted by Sertoli cells.   B. GnRH
57) Produced by the hypothalamus.   C. Inhibin
58) Released by the corpus luteum.   D. FSH
                                                E. Estrogen

    Answers: 55) D. 56) C. 57) B. 58) E.

59) Stimulates oogenesis.   A. ICSH
 0) Stimulates formation of testosterone.   B. FSH
) Released by the corpus luteum.   C. Relaxin
    Stimulates ejection of milk from   D. Oxytocin
    mammary glands.

                                                E. hCG

    nswers: 59) B. 60) A. 61) C. 62) D.

424

63) Tubing packed onto the posterior borders of the testis.

64) Pea-sized glands that produce mucus.

65) Inferior to the urinary bladder.

66) Ejaculatory duct.

A. Ductus deferens
B. Seminal vesicles
C. Bulbourethral gland
D. Prostate gland
E. Epididymis

Answers: 63) E. 64) C. 65) D. 66) A.

67) Stem cells.

68) Meiosis I forms.

69) Meiosis II forms.

70) Flagellum develops, resulting in

A. Primary spermatocyte
B. Secondary spermatocyte
C. Spermatogonium
D. Spermatids
E. Sperm cells

Answers: 67) C. 68) B. 69) D. 70) E.

SHORT ANSWER. Write the word or phrase that best completes each statement or answers the question.

71) The drug Viagra enhances the effects of _____ in the penis.

Answer: nitric oxide

72) The role of nitric oxide in the erection of the penis is to _____.

Answer: relax smooth muscle in penile arteries

73) The corpus luteum is maintained following fertilization by a hormone called _____.

Answer: human chorionic gonadotropin

74) Ovulation is triggered by a surge in the level of the hormone _____.

Answer: LH

75) High levels of _____ in the late preovulatory phase exert a positive feedback effect on LH and GnRH.

Answer: estrogens

76) During the ovarian preovulatory phase, the uterus is in its _____ phase in which endometrial mass doubles under the influence of _____.

Answer: proliferative; estrogens

77) A decrease in levels of progesterone stimulates constriction of spiral arterioles by stimulating the release of _____.

Answer: prostaglandins

78) Inhibin primarily inhibits secretion of _____ .

Answer: FSH

79) The three main estrogens are _____ , _____ , and _____ .

Answer: beta-estradiol; estrone; estriol

80) Milk is stored near the nipples in spaces called _____ .

Answer: lactiferous sinuses

81) Milk production is stimulated mainly by the hormone _____ , and ejection of milk is stimulated mainly by the hormone _____ .

Answer: prolactin; oxytocin

82) _____ in seminal fluid and _____ in prostate fluid are used by sperm for ATP production.

Answer: Fructose; citric acid

83) About 60% of the volume of semen is contributed by the _____ .

Answer: seminal vesicles

84) The spermatic cord passes through an opening in the anterior abdominal wall called the _____ .

Answer: inguinal canal

85) Columnar cells in the epididymis have long, branching microvilli called _____ , which increase the surface area for _____ .

Answer: stercocilia; resorption of degenerated sperm

86) The source of androgen-binding protein is the _____ ; the function of androgen-binding protein is to _____ .

Answer: Sertoli cells; keep testosterone levels high near seminiferous tubules

87) In the prostate and seminal vesicles, 5 alpha-reductase converts testosterone to a more potent androgen called _____ .

Answer: dihydrotestosterone (DHT)

88) The hyaluronidase-containing vesicle of a sperm cell is called the _____ .

Answer: acrosome

89) The blood-testis barrier is formed just internal to the basement membrane of the seminiferous tubules by tight junctions between _____ cells.

Answer: Sertoli

90) The cells in the seminiferous tubules that secrete testosterone are the _____.

Answer: Leydig cells

91) Division of each primary spermatocyte eventually produces _____ spermatids.

Answer: four

92) The testicular artery, veins, autonomic nerves, lymphatic vessels, and the cremaster muscle together constitute the _____.

Answer: spermatic cord

93) The average volume of semen in an ejaculation is _____ with a sperm count of _____ per milliliter.

Answer: 2.5–5 mL; 50–150 million

94) The vascular changes resulting in an erection are the result of a _____ reflex.

Answer: parasympathetic

95) The foreskin of the penis is also known as the _____.

Answer: prepuce

ESSAY. Write your answer in the space provided or on a separate sheet of paper.

96) Describe the role of the autonomic nervous system in the human sexual response.

Answer: In arousal, the parasympathetic nervous system stimulates vasocongestion of sexual organs, secretion of lubricating fluids, and relaxation of vaginal smooth muscle. The sympathetic nervous system stimulates an increase in heart rate and force of contraction, an increase in vasomotor tone, and hyperventilation. During orgasm, the sympathetic nervous system stimulates rhythmic contractions of smooth muscle in genital organs, ejaculation of semen, and closing off of the male bladder.

97) Describe the functions of testosterone.

Answer: Testosterone promotes the development and maintenance of male secondary sex characteristics, protein anabolism, development of sexual function (behavior, libido, spermatogenesis), and the male pattern of development during prenatal life.

98) Compare and contrast the processes of spermatogenesis and oogenesis.

Answer: Both processes result in formation of gametes (haploid cells). One female stem cells yields one functional gamete plus 2–3 polar bodies; one male stem cell yields four functional gametes. Once spermatogenesis begins at puberty, it continues throughout life, producing 300 million sperm daily. Oogenesis begins during fetal life, resumes at puberty and produces one mature gamete monthly until menopause.

99) Describe the role of cervical mucus in the female reproductive tract.

Answer: Mucus forms a plug to impede penetration of sperm except at ovulation. It also supplements the energy needs of the sperm, serves as a sperm reservoir, and protects the sperm from pH damage and phagocytosis. Mucus also plays a role in capacitation.

100) Describe the positive feedback loop involved in ovulation.

Answer: FSH and LH promote follicular development, thus increasing estrogen production. High levels of estrogen during the late preovulatory phase stimulate release of GnRH from the hypothalamus. GnRH promotes release of more FSH and LH from the anterior pituitary.

CHAPTER 29   Development and Inheritance

MULTIPLE CHOICE. Choose the one alternative that best completes the statement or answers the question.

1) The term *capacitation* refers to
   A) union of the male and female pronuclei.
   B) equatorial division of the secondary oocyte following penetration by a sperm cell.
   C) functional changes that sperm undergo in the female reproductive tract that allow them to fertilize the secondary oocyte.
   D) functional changes in the zona pellucida caused by release of calcium ions.
   E) a sperm cell's penetration of the zona pellucida and entry into a secondary oocyte.

   Answer: C

2) What percentage of sperm cells introduced into the vagina normally reach the oocyte?
   A) less than 1%
   B) about 10%
   C) 25–30%
   D) 50%
   E) close to 100%

   Answer: A

3) The term *syngamy* refers to the
   A) union of male and female pronuclei.
   B) equatorial division of the secondary oocyte following penetration by a sperm cell.
   C) functional changes that sperm undergo in the female reproductive tract that allow them to fertilize a secondary oocyte.
   D) functional changes in the zona pellucida and entry into a secondary oocyte.
   E) a sperm cell's penetration of the zona pellucida and entry into a secondary oocyte.

   Answer: E

4) What is the next event following syngamy?
   A) penetration of the zona pellucida by a sperm cell
   B) depolarization and release of calcium ions by the oocyte
   C) cleavage
   D) meiosis II
   E) implantation

   Answer: B

5) How many days after fertilization does implantation of the blastocyst occur?
   A) 2 days
   B) 6 days
   C) 14 days
   D) 28 days
   E) 120 days

   Answer: B

6) At day 4 after fertilization, the solid ball of cells that has formed is called the
   A) zygote.
   B) blastocyst.
   C) gastrula.
   D) morula.
   E) embryo.

   Answer: D

7) Implantation usually occurs in the
   A) uterine tube.
   B) myometrium.
   C) cervix adjacent to the internal os.
   D) posterior fornix.
   E) posterior wall of the body or fundus of the uterus.

   Answer: E

8) The enzymes that allow implantation to occur are produced by the
   A) syncytiotrophoblast.
   B) cytotrophoblast.
   C) blastocele.
   D) inner cell mass.
   E) acrosome.

   Answer: A

9) The human gestation period is about
   A) 9 weeks.
   B) 24 weeks.
   C) 32 weeks.
   D) 38 weeks.
   E) 48 weeks.

   Answer: D

10) The chorion develops from the
    A) inner cell mass.
    B) blastocele.
    C) trophoblast.
    D) yolk sac.
    E) corpus luteum.
    Answer: C

11) The embryonic disc develops from the
    A) inner cell mass.
    B) blastocele.
    C) trophoblast.
    D) yolk sac.
    E) corpus luteum.
    Answer: A

12) The fetus is protected from mechanical injury by fluid contained within the
    A) allantois.
    B) yolk sac.
    C) blastocele.
    D) amnion.
    E) decidua.
    Answer: D

13) The embryonic period of development covers what time period?
    A) the time between fertilization and implantation
    B) the first two months following fertilization
    C) the first trimester of pregnancy
    D) the first two trimesters of pregnancy
    E) the time between the appearance
    Answer: B

14) For fertilization to occur, a sperm must penetrate
    A) corona radiata.
    B) zona pellucida.
    C) granulosa cells.
    D) clear glycoprotein layer.
    E) All answers are correct.
    Answer: E

15) The amnion forms from the
   A) syncytiotrophoblast.
   B) blastocele.
   C) cytrotrophoblast.
   D) decidua.
   E) inner cell mass.

   Answer: C

16) The yolk sac forms from the
   A) ectoderm of the inner cell mass.
   B) fluid within the amniotic cavity.
   C) allantois.
   D) decidua.
   E) endoderm of the inner cell mass.

   Answer: E

17) The extraembryonic coelom becomes the
   A) ectoderm.
   B) endoderm.
   C) amniotic cavity.
   D) ventral body cavity.
   E) blastocele.

   Answer: D

18) The "water" referred to when a woman's "water breaks" prior to delivery is
   A) amniotic fluid released when the amnion ruptures.
   B) maternal plasma leaking from weakened blood vessels.
   C) maternal urine expelled as compression of the bladder occurs.
   D) interstitial fluid released by separation of the layers of the placenta.
   E) mucus from stimulated glands in the cervix and vagina.

   Answer: A

19) Place the following terms into their correct time sequence:
   1) fertilization
   2) blastocyst
   3) cleavage
   4) zygote
   5) implantation
   6) development of chorionic villi
   A) 1, 2, 3, 4, 5, 6
   B) 1, 4, 3, 2, 5, 6
   C) 4, 5, 1, 2, 3, 6
   D) 3, 2, 1, 4, 6, 5
   E) 2, 1, 6, 5, 3, 4
   Answer: B

20) The endometrium is digested by enzymes released from the
   A) inner cell mass.
   B) chorionic villi.
   C) trophoblast.
   D) blastocoel.
   E) morula.
   Answer: C

21) Exchange of gases, nutrients, and wastes between maternal and fetal blood takes place between the
   A) amnion and chorion.
   B) decidua capsularis and chorion.
   C) decidua basalis and yolk sac.
   D) decidua basalis and chorionic villi.
   E) decidua capsularis and decidua basalis.
   Answer: D

22) The portion of the endometrium that becomes the maternal portion of the placenta is the
   A) decidua parietalis.
   B) decidua basalis.
   C) decidua capularis.
   D) stratum basalis.
   E) amnion.
   Answer: B

23) What fills intervillous spaces?
   A) amniotic fluid.
   B) maternal blood only.
   C) fetal blood only.
   D) interstitial fluid.
   E) both fetal and maternal blood.

   Answer: B

24) The portion of the endometrium that covers the embryo and is located between the embryo and the uterine cavity is the
   A) decidua parietalis.
   B) decidua basalis.
   C) decidua capsularis.
   D) stratum basalis.
   E) amnion.

   Answer: C

25) Most materials cross the placenta by
   A) primary active transport.
   B) filtration.
   C) phagocytosis.
   D) diffusion.
   E) exocytosis.

   Answer: D

26) Deoxygenated fetal blood is carried to the placenta via the
   A) uterine arteries.
   B) uterine veins.
   C) umbilical arteries.
   D) umbilical vein.
   E) decidua basalis.

   Answer: C

27) Which of the following has occurred by the end of the first month of development?
   A) The placenta has developed.
   B) The eyes have formed and are open.
   C) Ossification has begun.
   D) Urine has begun to form.
   E) The heart has formed and begun beating.

   Answer: E

28) The reason the corpus luteum is maintained in early pregnancy is to

   A) continue production of hCG.

   B) produce ATP for the developing embryo.

   C) keep levels of estrogen and progesterone high enough to maintain the endometrium.

   D) produce the enzymes necessary for implantation.

   E) serve as the connection between the developing embryo and the endometrium.

   Answer: C

29) Human chorionic gonadotropin is produced by the

   A) corpus luteum.

   B) embryo.

   C) endometrium.

   D) trophoblast cells of the chorion.

   E) blastocele.

   Answer: D

30) During early pregnancy, the main function of hCG is to

   A) nourish the embryo.

   B) stimulate placental growth.

   C) maintain the corpus luteum.

   D) depress estrogen production.

   E) prevent passage of testosterone to the developing embryo.

   Answer: C

31) Human chorionic gonadotropin is at its highest levels during

   A) ovulation.

   B) fertilization.

   C) implantation.

   D) the ninth week of pregnancy.

   E) the third month of pregnancy.

   Answer: D

32) Once hCG levels decrease, estrogen and progesterone are secreted mainly by the

   A) placenta.

   B) embryo.

   C) corpus luteum.

   D) hypothalamus.

   E) stratum basalis.

   Answer: A

33) Peak secretion of hCS occurs
    A) between fertilization and implantation.
    B) during the second trimester.
    C) late in the third trimester.
    D) just after delivery.
    E) at about the ninth week of pregnancy.

Answer: C

34) In a pregnant woman, decreased utilization of glucose and increased release of fatty acids from adipose tissue are promoted by the hormone
    A) estrogen.
    B) progesterone.
    C) inhibin.
    D) hCS.
    E) relaxin.

Answer: D

35) Early pregnancy tests are based on detection of what substance in the urine?
    A) amniotic fluid.
    B) blastomeres.
    C) hCG.
    D) high levels of progesterone.
    E) hCS.

Answer: C

36) What event marks the beginning of the stage of expulsion in true labor?
    A) when the woman's water breaks
    B) appearance of lochia
    C) complete cervical dilation
    D) severing of the umbilical cord
    E) completion of two hours of rhythmic uterine contractions

Answer: C

37) Colostrum is different from true milk because it contains less lactose and virtually no
    A) fat.
    B) protein.
    C) sodium.
    D) iron.
    E) antibodies.

Answer: A

38) A couple who are both phenotypically normal have a child who expresses a sex-linked recessive trait. Which of the following represents this child's genotype? [Let the trait be designated T (dominant) or t (recessive).]

A) Tt

B) tt

C) $X^t X^t$

D) $X^t Y$

E) Both C and D could be correct.

Answer: D

39) The observable characteristics of a person's genetic makeup are known as the

A) genotype.

B) phenotype.

C) gene pool.

D) karyotype.

E) autosomes.

Answer: B

40) An individual whose alleles for a particular trait are the same is said to be

A) dominant.

B) recessive.

C) homologous.

D) homozygous.

E) heterozygous.

Answer: D

41) The two alternative forms of a gene that code for the same trait and are at the same locus on homologous chromosomes are called

A) alleles.

B) autosomes.

C) Barr bodies.

D) blastomeres.

E) chromatids.

Answer: A

42) A person is heterozygous for a particular trait if he/she has
   A) more than two copies of a particular gene.
   B) genes for the trait on both sex chromosomes and autosomes.
   C) one dominant allele and one recessive allele for the trait.
   D) two dominant alleles for the trait.
   E) two recessive alleles for the trait.

   Answer: C

43) If a person's phenotype is intermediate between homozygous dominant and homozygous recessive then inheritance of this trait is an example of
   A) codominance.
   B) sex-linked inheritance.
   C) incomplete dominance.
   D) nondisjunction.
   E) lyonization.

   Answer: C

44) What would be the possible blood phenotypes of the offspring of parents who are genotype II and $I^A I^B$?
   A) type O only
   B) type AB only
   C) types A or B only
   D) types A, B, or O only
   E) All ABO types are possible.

   Answer: C

45) Inheritance of the ABO blood type is an example of
   A) codominance.
   B) sex-linked inheritance.
   C) incomplete dominance.
   D) nondisjunction.
   E) lyonization.

   Answer: A

46) How many pairs of autosomes does a normal human have?
   A) 22
   B) 23
   C) 44
   D) 46
   E) one

   Answer: A

47) A child expresses an autosomal recessive trait. Which of the following are NOT possible genotypes for the parents?

A) Both father and mother are homozygous recessive.

B) Both father and mother are heterozygous.

C) The father is heterozygous, and the mother is homozygous recessive.

D) The mother is heterozygous, and the father is homozygous recessive.

E) The father is heterozygous, and the mother is homozygous dominant.

Answer: E

48) By doing karyotyping, one can determine the gender of the child because

A) the external genitalia can be seen on the screen.

B) the total number of chromosomes would be different between the sexes.

C) the Y chromosome is much smaller than the X chromosome.

D) levels of testosterone are higher in the amniotic fluid of male embryos.

E) the embryonic membranes are thicker around male embryos.

Answer: C

49) Teratogens are

A) traits carried only on sex chromosomes.

B) agents that induce physical defects in developing embryos.

C) cells with an abnormal number of chromosomes.

D) all of the alleles that contribute to a particular trait.

E) genes that are inactivated during fetal development.

Answer: B

50) Most pregnancy tests are based on the presence or absence of

A) androgens.

B) progesterone.

C) estrogen.

D) human chorionic gonadotropin (hCG).

E) FSH and LH

Answer: D.

MATCHING. Choose the item in column 2 that best matches each item in column 1.

51) Inhibits uterine contractility.

52) Promotes milk production.

53) Stimulates uterine contractility.

54) Secreted by trophoblasts.

A. hCG

B. Progesterone

C. Oxytocin

D. Prolactin

E. Relaxin

Answers: 51) E. 52) D. 53) C. 54) A.

55) Solid mass of cells called blastomeres.   A. Zygote
56) Part of blastocyst that secretes hCG.   B. Trophoblast
57) Hollow ball of cells that implants into   C. Morula
    uterine wall.
58) Follows fertilization.   D. Blastocyst
                                               E. Amnion

Answers: 55) C. 56) B. 57) D. 58) A.

59) Heart forms.   A. 1–4 weeks of gestation
60) Primary brain vesicles develop into   B. 5–8 weeks
    secondary brain vesicles.
61) External ears develop.   C. 9–12 weeks
62) Fetal moves are commonly felt by mother.   D. 13–16 weeks
                                               E. 17–20 weeks

Answers: 59) A. 60) B. 61) C. 62) E.

63) Epithelial lining of the gastrointestinal. A. Ectoderm
64) Cardiac muscle tissue.   B. Mesoderm
65) All nervous tissue.   C. Endoderm
66) Stem cells.   D. All germ layers are correct

Answers: 63) C. 64) B. 65) A. 66) D.

67) Refers to how the genetic makeup is expressed. A. Phenotype
68) Alleles of a gene pair are identical.   B. Homozygous
69) Non-sex chromosomes.   C. Autosomes
70) Alleles of gene pair are nonidentical.   D. Heterozygous
                                               E. Genotype

Answers: 67) A. 68) B. 69) C. 70) D.

SHORT ANSWER. Write the word or phrase that best completes each statement or answers the question.

71) Fusion of a sperm with a secondary oocyte is called _____.

Answer: syngamy

72) Smaller cells produced by cleavage of the zygote are called _____.

Answer: blastomeres

73) In the region of contact between the blastocyst and the endometrium, the trophoblast develops two layers: the _____ and the _____.

Answer: syncytiotrophoblast; cytotrophoblast

74) The inner cell mass is called the _____ once the amniotic cavity develops.

Answer: embryonic disc

75) The organ that is the site of exchange of nutrients and wastes between mother and fetus is the _____.

Answer: placenta

76) During the first three to four months of pregnancy, the lining of the uterus is maintained by progesterone and estrogen secreted by the _____.

Answer: corpus luteum

77) Early pregnancy tests detect the hormone _____.

Answer: hCG

78) A cell that has one or more chromosomes of a set added or deleted is called _____.

Answer: aneuploid

79) The prime male-determining gene is called _____ and is located on the _____ chromosome.

Answer: SRY; Y

80) An agent that causes developmental defects in the embryo is called a(n) _____.

Answer: teratogen

81) If two genes for a trait are expressed equally in a heterozygote, they are said to be _____.

Answer: codominant

82) A permanent, heritable change in a gene that causes it to have a different effect than it previously had is called a(n) _____.

Answer: mutation

83) The phenomenon in which the phenotype is dramatically different depending on the parental origin is called _____.

Answer: genomic imprinting

84) At parturition the androgen _____ is secreted from the fetal adrenal cortex and converted to _____ by the placenta.

Answer: DHEA; estrogen

85) The hormone _____ inhibits uterine contractions; the hormone _____ promotes uterine contractions.

Answer: progesterone; oxytocin

86) From the third to the ninth month of pregnancy, estrogen and progesterone levels are kept high enough to maintain the pregnancy by the _____.

Answer: placenta

87) The glycoprotein receptor for sperm called ZP3 is located in a glycoprotein layer around the oocytes called the _____.

Answer: zona pellucida

88) _____ is the term for the functional changes that sperm undergo in the female reproductive tract that allow them to fertilize a secondary oocyte.

Answer: Capacitation

89) The fertilized ovum is called a(n) _____.

Answer: zygote

90) The part of the female reproductive tract in which fertilization normally occurs is the _____.

Answer: uterine tube

91) By the end of the third day after fertilization, the fertilized egg has become a solid ball of cells called the _____.

Answer: morula

92) The hollow ball of cells that is implanted into the uterine wall is called the _____.

Answer: blastocyst

93) The primary germ layers are the _____, the _____, and the _____.

Answer: ectoderm; mesoderm; endoderm

94) The structure derived from the trophoblast of the blastocyst that becomes the principal embryonic part of the placenta is the _____.

Answer: chorion

95) The chorion of the placenta secretes the hormone _____, which mimics the action of LH.

Answer: human chorionic gonadotropin

ESSAY. Write your answer in the space provided or on a separate sheet of paper.

96) Describe the composition and functions of amniotic fluid.

Answer: Initially, amniotic fluid is a filtrate of maternal blood; then fetal urine is added daily. It acts as a shock absorber and fetal temperature regulator, and it prevents adhesion of fetal skin to surrounding tissues.

97) Describe the process and purpose of amniocentesis.

Answer: The position of the fetus and placenta is identified via ultrasound and palpation, and the skin is prepared with antiseptic and local anesthetic. A hypodermic needle is inserted through the abdominal wall and uterus to withdraw 10 mL of amniotic fluid from the amniotic cavity. The fluid and cells are examined and biochemically tested for abnormal proteins and chromosome abnormalities that may signal fetal problems and congenital defects.

98) Describe the hormonal events surrounding parturition.

Answer: Fetal CRH secretion increases, which causes estrogen to increase as fetal ACTH triggers an increase in cortisol and DHEA, which is converted to estrogen by the placenta. Estrogen increases oxytocin receptors on uterine smooth muscle fibers and makes them form gap junctions. Oxytocin stimulates uterine contraction, and relaxin dilates the cervix and loosens the pubic symphysis. Estrogen also increases prostaglandins to digest collagen in the cervix. Oxytocin the cervix, and the hypothalamus maintain a positive feedback loop to maintain labor.

99) Describe the potential hazards to the embryo and fetus associated with alcohol consumption and cigarette smoking.

Answer: Acetaldehyde, a metabolic product of alcohol, causes fetal alcohol syndrome, which is characterized by slow growth, small head, unusual facial features, defective heart and other organs, malformed limbs, and CNS abnormalities that may lead to behavioral problems. Cigarette smoking leads to low birth weight and increased risk of fetal/infant mortality, cardiac problems, anencephaly, and cleft lip and cleft palate.

100) Describe the events of early development from fertilization through implantation.

Answer: At fertilization, the male and female pronuclei fuse into the segmentation nucleus, which, with cytoplasm and zona pellucida, form the zygote in the uterine tube. The zygote undergoes cleavage to form blastomeres, which are arranged as a solid ball of cells called the morula. By the fifth day after fertilization, the hollow blastocyst has formed (trophoblast, inner cell mass, and blastocele), which implants into the uterine wall oriented toward the endometrium.